GEORGE BERKELEY: RELIGION AND SCIENCE IN THE AGE OF ENLIGHTENMENT

ARCHIVES INTERNATIONALES D'HISTOIRE DES IDÉES

INTERNATIONAL ARCHIVES OF THE HISTORY OF IDEAS

201

GEORGE BERKELEY: RELIGION AND SCIENCE IN THE AGE OF ENLIGHTENMENT

Silvia Parigi

George Berkeley: Religion and Science in the Age of Enlightenment

Silvia Parigi
Università di Cassino

Springer

Editor
Silvia Parigi
Università di Cassino
Fac. Lettere e Filosofia
Via Zamosch 1
03043 Cassino
Italy
Silpari@libero.it

ISBN 978-90-481-9242-7 e-ISBN 978-90-481-9243-4
DOI 10.1007/978-90-481-9243-4
Springer Dordrecht Heidelberg London New York

Library of Congress Control Number: 2010935167

Printed on acid-free paper

Springer is part of Springer Science+Business Media (www.springer.com)

Acknowledgments

Most of the essays collected here were presented, in a first version, as contributions to the International Berkeley Conference that took place in Gaeta (Italy) in September 2007, supported by the Philosophy and Communication Department of the University of Cassino, the Metanexus Institute of Philadelphia and the International Berkeley Society. In addition to these institutions, I would like to thank the many distinguished scholars who came from Sweden, Estonia, Switzerland, France, and even from Canada and the United States of America: Bertil Belfrage, Costica Bradatan, Richard Brook, Geneviève Brykman, C. George Caffentzis, Sébastien Charles, Stephen H. Daniel, Marc A. Hight, Roomet Jakapi, Luc Peterschmitt, Claire Schwartz. I would also like to thank the Italian scholars who took part in the Conference: Daniele Bertini, Antonino Drago, Caterina Menichelli, Massimo Stanzione, and Eva Valeriani for her precious organizational work. All of these people did not limit themselves to giving papers and discussing them, but made it possible, for three days, to create an authentic learning community, sharing interests and passion for knowledge.

I wish also to thank Peter Kail and David Berman for their interest in this initiative and their contributions to the present collection, and Siobhan Nott for her competence and patience in the linguistic revision.

A special thanks is due to Antonio Clericuzio, who made the Conference possible, and enriched it with his comments and suggestions.

This book is dedicated to the memory of my father.

Contents

Introduction

Berkeley's Philosophy Between the Analytics and the Historians: Beyond the "Standard Interpretation"

Up to now, the critical approaches of the scholars towards Berkeley's philosophy may be summarized as follows: on the one hand, the analytical attitude, dominant in the 1970s–1980s of the last century in English-speaking countries, focused on the early published works (above all *Treatise on the Principles of Human Knowledge* and *Three Dialogues between Hylas and Philonous*), considered as originally meaningful and eminently significant.[1] Interpreters found some problems of "consistency" in Berkeley's texts, but the constant endeavour *to give them sense* was usually successful: they were able to resolve problems, absorbing – so to say – Berkeley's "inconsistencies". Moreover, analytic scholars often judged Berkeley's philosophy from the point of view of the (then) current philosophical theories, considered as the worthiest to be taken into consideration.

On the other hand, historians of philosophy – more often "continental" or Irish – dedicated themselves to the whole of Berkeley's life and works, including the less palatable ones, either because unpublished (as *Philosophical Commentaries*), or

---

[1]See, for example, Jonathan Bennett, *Locke, Berkeley, Hume. Central Themes* (Oxford: Clarendon Press, 1971); Ian C. Tipton, *Berkeley. The Philosophy of Immaterialism* (London: Methuen, 1974); re-edited in *The Philosophy of George Berkeley*, ed. George Pitcher (Garland: New York and London, 1988–1989); Geoffrey J. Warnock, *Berkeley* (Oxford: Blackwell, 1982); J. O. Urmson, *Berkeley* (Oxford: Oxford University Press, 1982); George Pitcher, *Berkeley* (London: Routledge, 1984); Noel Fleming, "Berkeley and Idealism", *Philosophy*, 60 (1985): 309–325; Anthony C. Grayling, *Berkeley: The Central Arguments* (London: Duckworth, 1986); Jonathan Dancy, *Berkeley: An Introduction* (Oxford: Blackwell, 1987); P.D. Cummins, "On the Status of Visuals in Berkeley's New Theory of Vision", in *Essays on the Philosophy of George Berkeley*, ed. Ernest Sosa (Dordrecht: Reidel, 1987), 165–194; Daniel Flage, *Berkeley's Doctrine of Notions* (London: Croom Helm, 1987), Kenneth P. Winkler, *Berkeley: An Interpretation* (Oxford: Clarendon Press, 1989). See also the more recent contribution by P. D. Cummins, "Berkeley on Mind and Agency", in *The Cambridge Companion to Berkeley*, ed. Kenneth P. Winkler (New York: Cambridge University Press, 2005), 209–218.

because someway "deviating" from the mainstream of Berkeley's thought (as the essays published in the *Guardian*, *Alciphron* and, above all, *Siris*).[2] Moreover, historians tried to comprehend Berkeley's life and works, tracing them back to their historical context; they even accepted the possible "inconsistencies", and considered them not as unpleasant and embarrassing contradictions to eliminate, but as

[2]Cf. John Wild, *George Berkeley: A Study of His Life and Philosophy* (Cambrige, MA: Harvard University Press, 1936); A.A. Luce, *The Life of George Berkeley* (Edinburgh and London: Nelson, 1949); Martial Gueroult, *Berkeley. Quatre études sur la perception et sur Dieu* (Aubier: Montaigne, 1956); André-Louis Leroy, *George Berkeley* (Paris: PUF, 1959); Geneviève Brykman, *Berkeley. Philosophie et apologétique* (Paris: Vrin, 1984); David Berman, *George Berkeley. Idealism and the Man* (Oxford: Clarendon Press, 1994); Dominique Berlioz, *Berkeley: un nominalisme réaliste* (Paris: Vrin, 2000). On *Philosophical Commentaries*, see A.A. Luce, "Berkeley's Commonplace Book. Its Date, Purpose, Structure and Marginal Signs", *Hermathena* 22 (1931): 99–131; Bertil Belfrage, "The Order and Dating of Berkeley's Notebooks", *Revue internationale de philosophie*, 154 (1985): 196–214; "The Clash on Semantics in Berkeley's Notebook A", in *George Berkeley. Essays and Replies*, ed. David Berman (Dublin: Irish Academic Press in association with Hermathena, 1986), 117–126; Robert McKim, "The Entries in Berkeley's Notebooks: A Reply to Bertil Belfrage", ivi, 156–161; Bertil Belfrage, "A New Approach to Berkeley's Philosophical Notebooks", in Sosa, *Essays on the Philosophy of George Berkeley*, 217–230. On the *Guardian essays*, see David Berman, "Did Berkeley write *Guardian* 130?," *Berkeley Newsletter* 5 (1981): 10–13; J. Stephens (ed.), *The Guardian* (Kentucky: The University Press, 1982); David Berman, *Review of The Guardian, ed.* J. C. Stephens, *Berkeley Newsletter* 7 (1984): 23–26; Kenneth P. Winkler, "The Authorship of *Guardian* 69," *Berkeley Newsletter* 7 (1984): 1–6. On *Alciphron*, cf. Paul Olscamp, *The Moral Philosophy of George Berkeley* (The Hague: Nijhoff, 1970); David Berman, "Cognitive Theology and Emotive Mysteries in Berkeley's *Alciphron*", *Proceedings of the Royal Irish Academy* (1981): 219–229; *Alciphron*, or *the Minute Philosopher: in focus*, ed. David Berman (London; New York: Routledge, 1993); Roomet Jakapi, "Emotive Meaning and Christian Mysteries in Berkeley's *Alciphron*", *British Journal for the History of Philosophy* 10 (2002): 401–411; "Faith, Truth, Revelation and Meaning in Berkeley's Defence of the Christian Religion (in *Alciphron*)", *The Modern Schoolman* 80 (2002): 23–34; Costica Bradatan, "Rhetoric of Faith and Patterns of Persuasion in Berkeley's *Alciphron*", *The Heythrop Journal* 47 (2006): 544–561. On *Siris*, see N. Baladi, "Plotin et l'immatérialisme de Berkeley: témoignage de la *Siris*," in *Plotino e il neoplatonismo in Oriente e in Occidente*, (Roma: Accademia Nazionale dei Lincei, 1974): 597–604; Gabriel Moked, "Two Central Issues in Bishop Berkeley's 'Corpuscularian Philosophy' in the *Siris*", *History of European Ideas* 7 (1986): 643–649; J. O. Urmson, "Berkeley's Philosophy of Science in the *Siris*," *History of European Ideas* 7 (1986): 562–566; Luigi Neri, "Il caso *Siris*, ovvero la 'seconda filosofia' di George Berkeley vescovo di Cloyne," *Giornale critico della filosofia italiana* (1990): 320–341; Marina Benjamin, "Medicine, Morality and the Politics of Berkeley's Tar-Water", in *The Medical Enlightenment of the Eighteenth Century*, eds. A. Cunningham and R. French (Cambridge: Cambridge University Press, 1990), 165–193; Lisa J. Downing, "*Siris* and the Scope of Berkeley's Instrumentalism," *British Journal for the History of Philosophy* 3 (1995): 279–300; Sébastien Charles, "The *Siris* in the Age of Enlightenment: Panacea or Imposture?," *Hermathena* 168 (2000): 55–69; Costica Bradatan, "'One is All, and All is One'. The Great Chain of Being in Berkeley's *Siris*", in F. O'Gorman and D. Donald (eds.), *Ordering the World in the Eighteenth Century* (Palgrave Macmillan, 2005), 63–82; Timo Airaksinen, "The Chain and the Animal: Idealism in Berkeley's *Siris*", in *Eriugena, Berkeley, and the Idealist Tradition*, eds. Stephen Gersh and Dermot Moran (Notre Dame, IN: University of Notre Dame Press, 2006), 224–243 and "The Path of Fire: the Meaning and Interpretation of Berkeley's *Siris*", in *New Interpretations of Berkeley's Thought*, ed. Stephen H. Daniel (Amherst, NY: Humanity Books, 2007), 261–281.

possible changes in opinion, or elements of tenseness to account for from an historical point of view.

From this perspective, it was not always possible *to solve* interpretive problems; perhaps simply because Berkeley himself left them unsolved, or because they were the symptoms of some yet unexplored historical "knots". To sum up, historians are used to cohabiting with contradictions, while analytics are determined to accurately sort out and denounce them. Or, to put it in other terms, historians do not need to be "charitable", while analytics often recur to this word, in order to express their attitude towards leading philosophical figures of the past, and their texts.

Last of all, historians of science generally overlooked Berkeley's "scientific" works (as *New Theory of Vision, De motu, Analyst, Querist*, and – once again – *Siris*), maybe because they (and their author) were too philosophically compromised with the embarrassing metaphysical tenets of immaterialism.[3]

Sometimes, representative scholars of the main analytic philosophy reached conclusions that risk appearing quite obvious from an historical point of view, as if they amounted to saying that the philosophy of the twentieth century is different from the philosophy of the eighteenth. Other times, historical reconstructions did not fail to appear quite useless, from a theoretical point of view, either because they limited themselves to a general survey of Berkeley's life and works, or because they

[3]Significantly, the majority of studies on Berkeley's contributions to the sciences has been written by analytic philosophers or historians of philosophy: see J.O. Wisdom, "The Analyst Controversy: Berkeley's Influence on the Development of Mathematics", *Hermathena* 29 (1939): 3–29; "Berkeley's Criticism of the Infinitesimal," *British Journal for the Philosophy of Science*, 4 (1953): 22–25; P. Devaux, "Berkeley et les mathématiques", *Revue internationale de philosophie*, 7 (1953): 101–133; Thomas E. Jessop, *"Berkeley and Contemporary Physics"*, ivi, 87–100; Karl R. Popper, "A Note on Berkeley as a Precursor of Mach and Einstein", *British Journal for the Philosophy of Science*, 4 (1953): 26–36; G.J. Withrow, "Berkeley's Philosophy of Motion", *British Journal for the Philosophy of Science*" 4 (1953): 37–45; W.A. Suchting, "Berkeley's Criticism of Newton on Space and Time", *Isis*, 58 (1967): 186–197; Richard J. Brook, *Berkeley's Philosophy of Science* (The Hague: Nijhoff, 1973); M. Blay, "Deux moments de la critique du calcul infinitésimal: Michel Rolle et George Berkeley", *Revue d'Histoire des sciences* 39 (1983): 223–253; J. O. Urmson, "Berkeley's Philosophy of Science in the *Siris*", *History of European Ideas* 7 (1986): 562–66; George Pappas, "Science and Metaphysics in Berkeley", *International Studies in the Philosophy of Science* 2 (1987): 105–114; D. Sherry, "The Wake of Berkeley's *Analyst: rigor mathematicae*?", *Studies in History and Philosophy of Science* 18 (1987): 455–480; Gabriel Moked, *Particles and Ideas. Bishop Berkeley's Corpuscolarian Philosophy* (Oxford: Clarendon Press, 1988); Luigi Neri, *George Berkeley. Filosofia e critica dei linguaggi scientifici* (Bologna; CLUEB, 1990); *Margaret Atherton, Berkeley's Revolution in Vision* (Ithaca: Cornell University Press, 1990); Silvia Parigi, "I filosofi e il microscopio: da Descartes a Berkeley", *Rivista di storia della scienza*, 1 (1993): 155–172; Douglas M. Jesseph, *Berkeley's Philosophy of Mathematics* (Chicago, IL: University of Chicago Press, 1993); Silvia Parigi, *Il mondo visibile. George Berkeley e la "perspectiva"* (Firenze: Olschki, 1995); Sébastien Charles (ed.), *Science et épistémologie selon Berkeley* (Saint-Nicolas, Québec: Les Presses l'Université Laval, 2004); Luc Peterschmitt, *Sciences de la nature et philosophie dans la pensée de Berkeley* (Ph.D. dissertation, University of Lille 3, 2005).

insisted on his well-known relationships with Locke, and – after A. A. Luce's book[4] – with Malebranche, or his "influence" on Hume.[5]

The "standard approach" – as Stephen Daniel calls it in the opening essay of this collection – is therefore, in my opinion, a double one: the analytical approach consists of a quite exclusive consideration of *Treatise on the Principles of Human Knowledge* and *Three Dialogues between Hylas and Philonous*, taken out of their authentic historical context, and judged from the point of view of their *actual truth*. The historical approach includes Berkeley's thought in the empirical tradition (something like the old Hegelian triad: Locke – Berkeley – Hume), or, less obviously, in a Cartesian and Malebranchian context – are they to be considered as opposite to each other, or not.[6]

There is another limit – evident, but still unstressed – that is to be imputed to the "standard approach", either analytical or historical: scholars definitely privileged English-written critical literature, overlooking contributions written in other languages. That is to say, they tended to consider unimportant whatever was unintelligible to an English-speaking public.

In the last times, however, things have partially changed, in that Berkeley's philosophy has received more attention in its "psychological, experiential and observational" aspects, rather than in its "conceptual, analytic, and argumentative" elements, almost exclusively focused on by interpreters belonging to the "Anglo-American analytic tradition".[7] It has also been considered from the perspective of *its past*,

---

[4]A. A. Luce, *Berkeley and Malebranche. A Study in the Origins of Berkeley's Thought* (Oxford: Clarendon Press, 1934; second edition Oxford: Oxford University Press, 1967, then in Pitcher, *The Philosophy of George Berkeley*). See also Charles McCracken, *Malebranche and British Philosophy* (Oxford: Clarendon Press, 1983).

[5]R. Reininger, *Locke, Berkeley, Hume*, Reprint d. Ausg. München: E. Reinhardt, 1922 (Nendeln/Liechtenstein: Kraus, 1973); A. A. Luce, *The Dialectic of Immaterialism. An Account of the Making of Berkeley's Principles* (London: Hodder and Stoughton, 1963); Harry M. Bracken, "Bayle, Berkeley, and Hume," *Eighteenth-Century Studies* 11 (1977): 227–245; David Raynor, "Minima Sensibilia in Berkeley and Hume," *Dialogue* 19 (1980): 196–200; Ram Adhar Mall, *Der operative Begriff des Geistes: Locke, Berkeley, Hume* (Freiburg, Breisgau: Alber, 1984); Colin M. Turbayne, "Hume's Influence on Berkeley," *Revue internationale de philosophie* 154 (1985): 259–269; D.E. Bradshaw, "Berkeley and Hume on Abstraction and Generalization," *History of Philosophy Quarterly* 5 (1988): 11–22; R. Fogelin, "Hume and Berkeley on the Proofs of Infinite Divisibility," *Philosophical Review* 97 (1988): 47–69; David Raynor, "Hume and Berkeley's *Three Dialogues*", in *Studies in the Philosophy of the Scottish Enlightenment* (Oxford: Clarendon Press, 1990): 231–250; Margaret Atherton, *The Empiricists: Critical Essays on Locke, Berkeley, and Hume* (Lanham: Rowman & Littlefield, 1999); Daniele Bertini, "Hume e l'immaterialismo", *Aquinas* 49 (2006): 621–635.

[6]Charles McCracken singles out a Cartesian way to immaterialism, starting from Descartes and coming to Arthur Collier through Malebranche and John Norris ("Stages on a Cartesian Road to Immaterialism", *Journal of the History of Philosophy* 24 (1986): 19–40), while Silvia Parigi argues that there is a unique road, both empiristic and Cartesian ("Is there a Cartesian road to Immaterialism?", in *Berkeley et le cartésianisme*, ed. Geneviève Brykman (Nanterre, Université Paris X, 1997), 23–48).

[7]David Berman, Introduction to *Berkeley and Irish Philosophy* (London–New York: Continuum, 2005), 4.

rather than of *its future*,[8] in some unusual topics and from different points of view.[9] Analytical philosophers have begun to realize "the obvious problem of anachronism" (as P.J.E. Kail expresses himself, in the essay included in this volume) in their interpretive approach, while historians have not avoided facing the theoretical questions offered by Berkeley's thought. They have also widened their usual chronological limits, considering Berkeley's philosophy within Enlightenment(s),[10] comparing his thought with some unusual contemporaries',[11] or placing not only his idealism,[12] but his theory of vision and "medicine" as well,[13] in a more ancient tradition.

Moreover, it may be suspected that the subtle analytical discussions on particular epistemological and metaphysical topics did eventually prove less palatable to philosophical readers, than an honest analysis of Berkeley's works, taken in their chronological order and included in their contemporary philosophical, historical, political, theological and literary context.

Paraphrasing the very well-known Kantian sentence, we might affirm that philosophy without history is empty, while history without philosophy is blind: this is exactly the perspective that informs the present collection of essays, written by some leading European and American scholars. It focuses on an outstanding figure of eighteenth-century philosophical, scientifical and theological thought, who deserves to be explored in all his interests and concerns, in the details of his many-sided life and works.

George Berkeley was in fact considered "the most engaging and useful man in Ireland in the eighteenth century".[14] This hyperbolic statement refers both to

---

[8]Costica Bradatan, Introduction to *The Other Bishop Berkeley. An Exercise in Re-Enchantment* (New York: Fordham University Press, 2006), 1–17.

[9]For example, Geneviève Brykman stressed the importance of the "discursive context" (as Sthephen Daniel calls it in his essay, included in the present collection) in her book: *Berkeley et le voile des mots* (Paris: Vrin, 1993).

[10]Sébastien Charles, *Berkeley au siècle des Lumières* (Paris:Vrin, 2003) and "Berkeley and the *Lumières*: Misconception and Reconstruction", in Daniel, *New Interpretations of Berkeley's Thought*, 283–309. Stephen H. Daniel (New York: Humanity Books, 2008), 283–309.

[11]As to a contemporary outline of Berkeley's metaphysics and theories of ideas, see Parigi, *Il mondo visibile. George Berkeley e la "perspectiva"*, chap. 2; Richard Glauser, *Berkeley et les philosophes du XVIIe siècle* (Sprimont: Mardaga, 1999). See also the most recent study by Talia M. Bettcher, *Berkeley's Philosophy of Spirit* (London-New York: Continuum, 2007). A quite unusual relationship is drawn by Lawrence E. Klein, "Berkeley, Shaftesbury and the Meaning of Politeness", *Studies in Eighteenth-Century Culture* 16 (1986): 57–68.

[12]Gersh and Moran, *Eriugena, Berkeley and the Idealist Tradition.*

[13]As to the theory of vision, see Parigi, *Il mondo visibile. George Berkeley e la "perspectiva"*, chap. 1, where Berkeley's theory of vision is put against the background of medieval "perspectiva", starting from Alhazen and ending in Molyneux's *Dioptrica nova*. As to tar-water, see Benjamin, "Medicine, Morality and the Politics of Berkeley's Tar-Water"; Charles, "The *Siris* in the Age of Enlightenment"; Parigi, Introduzione to George Berkeley, *Opere filosofiche* (Torino: UTET, 1996), 9–51, where the origins of tar-water are traced back to Greek, Arabian, medieval and Renaissance medicine.

[14]B. Chance, "George Berkeley and *An Essay on Vision*", *Archives of Ophthalmology* 29 (1943): 605–614.

Berkeley's life and thought; in fact, he always considered himself a pioneer called to think and do new things. He was an empiricist well versed in the sciences, an amateur of the mechanical arts,[15] as well as a metaphysician; he was the author of many completely different discoveries (from the "new" theory of vision to the *esse est percipi*, from a "new" sensible geometry to tar-water), as well as a very active Christian, a zealous bishop and the apostle of the Bermuda project.

The previously unpublished essays collected in this volume aim to reconstruct the *complexity* of Berkeley's figure, without selecting "major" works, nor searching for "coherence" at any cost. They focus on different aspects of Berkeley's thought, showing their intersections; they explore the important contributions he gave to various scientific disciplines, as well as to the eighteenth-century philosophical and theological debate. They highlight, too, the wide influence that his presently most neglected or puzzling books had at that time; they avoid any anachronistical trial of Berkeley's thought, judged from a contemporary point of view, in order to state whether what he maintained *was* (or rather *is*) *right* or *wrong*.

As Daniel claims, it is no longer the case to refuse, as confused and contradictory, whatever is simply difficult to interpret from our point of view, or cannot be constrained in the "Cartesian and Lockean framework". It is not even proper to suppose changes, turning points and inconsistencies where a more accurate textual analysis and historical comprehension is just needed. Sure, it is easier to dismiss as not significant and "strange" those theories ("the language of nature, the semiotic character of things, spiritual substances, cosmic fire") and works (*Alciphron*, *Analyst*, *Querist*, and, above all, *Siris*) that cannot be interpreted in Cartesian or Lockean terms. But, as Luc Peterschmitt efficaciously suggests, this is only "the effect of our ignorance". In fact, it may be more comfortable to forget the context in which Berkeley lived and wrote, than to remember that he was a learned scholar, able to read Latin, Greek, Hebrew, Italian, Spanish and French, familiar with ancient and Renaissance sources. This is particularly important in the case of *Siris*, a work that has never ceased to puzzle scholars, because of its oddity and apparent lack of a causal order.

It may appear quite obvious that a volume collecting different contributions is not written from a unique, or prevailing, point of view: nevertheless, in my opinion there is a link among the distinguished Berkeleian scholars' essays collected here: that is to say, a clear consciousness of the insufficiency of the "standard approach" – whether analytical or historical – and a peculiar attention to the history of ideas, in that it may make a deep exploration of the relationship between philosophy, science and religion in Berkeley's works possible (also through the reading of non English-written books).

In fact, the essays that compose this volume suggest new interpretations of Berkeley's thought (Daniel, Hight, Kail, Berman, Schwartz, Parigi), as a whole (Daniel), or as

[15] Cf. S.R.L. Clark, "God-Appointed Berkeley and General Good", in *Essays on Berkeley,* ed. J. Foster and H. Robinson (Oxford: Clarendon Press, 1985), 233–253.

regards certain topics; they focus on usually neglected works (Caffentzis, Airaksinen, Bertini, Peterschmitt, Parigi), or propose an historical reading able to widen the common Cartesian-Lockean-Malebranchian perspective (Brykman, Charles, Menichelli, Berman, Bertini, Schwartz, Airaksinen, Peterschmitt, Parigi).

The final outcome is a less pacific portrait of the "good Bishop": for example, he was fascinated by Spinoza's powerfully heretical thought (Brykman), and was in fact accused of spinozism and atheism by some of his contemporaries (Menichelli); but, at the same time, he was an adversary of that movement of thought that Jonathan Israel has recently called "Radical Enlightenment" (Brykman). He could be considered as the exponent of a theology based on personal, sensible and emotive experience – in a word, on a kind of non-rational experience usually called "faith" (Bertini) – or, on the contrary, he should be seen as the proposer of a rational religion, based on philosophical reasoning (Berman). He was the follower of Greek and Renaissance theory of the cosmic spirit, and at the same time of Newton's anti-mechanistic doctrine of aether (Parigi).

According to the interpretive perspective outlined in this introduction, the essays that compose the present volume have been divided into three groups: those proposing new suggestions as to the meaning of Berkeley's thought; those focusing on some neglected Berkeleian works and aspects of the Bishop's thought; and those aimed at widening and clarifying the historical context in which Berkeley's works have to be placed. Nevertheless, many essays could have been included in more than one group.

In order to achieve a correct interpretation of Berkeley's works, Stephen Daniel proposes to simply follow Berkeley's own suggestion, stated in one of his letters to Samuel Johnson: to read his works in their chronological order. He shows the unity and coherence of Berkeley's philosophy both regarding the conception of bodies as powers (in his youthful notebooks) and the conception of mind as a created active being. Its aim should not be to gain a "homogenous unity", but "to achieve harmony in an ever-increasing variety of expressions"; not "to remove differences", but "to create multiplicity" – in accordance (I would add) with the Baconian tradition.[16]

In the second essay, Marc Hight claims, too, the unity and consistency of Berkeley's epistemology as regards a particular version of instrumentalism, that he thoroughly examines. Hight agrees with Daniel in stressing the importance of Berkeley's doctrine of signs: scientific concepts are *signs* (i.e. *instruments*), able and useful in explaining and organizing phenomenal experiences, in their variety and regularities.

Also Peter Kail – who analyses and compares Berkeley's and Hume's treatment of causal relation, from the point of view of some current perspectives in philosophy of language – argues that all the causal relations among ideas should be substituted, in Berkeley's opinion, by conventional sign/signifier relations.

---

[16]*In Novum Organum* (I, 112), Francis Bacon wrote: "Interim particularium multitudinem nemo reformidet, quin potius hoc ipsum ad spem revocet" ("In the meantime, let nobody fear the multiplicity of particulars; on the contrary, let it be a reason of hope"; translation mine).

The second section of the present collection groups different papers, dealing with some rather neglected topics and works. Claire Schwartz argues that Berkeley's philosophy of mathematics should not be interpreted as a "formalist" one, at least in the sense currently attributed to that word, which is deeply connected with an instrumentalist conception of truth. That is to say, it is not correct to apply the historiographically uncomfortable label of "precursor" to Berkeley. Some of his well-known contemporaries – Descartes, Malebranche, and in particular Leibniz – were "more in advance with the notion of a free constitution of formal systems", in consequence of their ontology, though it would not be proper to use the category of "formalism" as regards their philosophies of mathematics as well. Once again, from Schwartz's essay the importance of Berkeley's doctrine of signs stands out, as to the domain of mathematical thought.

George Caffentzis proposes an inhedit reassessment of the Hegelian triad: Locke, Berkeley and Hume are re-examined as philosophers of money, after answering the following questions: what does "philosopher of money" mean? Which conditions are to be fulfilled, in order to be considered a philosopher of money? What are the relationships between Locke, Berkeley and Hume's philosophies of money and their philosophies (i.e. ontologies and epistemologies, theories of ideas and mind)? Textual evidence is drawn from *Querist*, "a supposedly marginal text of eighty pages consisting all and only of questions", on whose importance Caffentzis began to bet nearly ten years ago. Moreover, he stresses the necessity of an historical approach to the philosophy of money, not forgetting the fact that "economics did not exist in its contemporary meaning until the later part of the nineteenth century".

Among Berkeley's puzzling and embarrassing works, *Siris* certainly holds the supremacy: it is not surprising, therefore, that it has long been neglected by scholars, who for many years spoke of Berkeley's "second" philosophy. In partial satisfaction of that, three essays deal with *Siris*, in the second section of the present volume. Luc Peterschmitt focuses on Section 202, questioning Berkeley's identification of Newton's acid with Homberg's sulphur, and of them both with Newton's aether, and Berkeley's aether or fire. In Peterschmitt's opinion, the effort Berkeley made to build a unified chemical theory (that *did not* actually *exist* in 1744) had the purpose to give a "convincing basis" to his "natural theology, starting from tar-water to trace back the chain of beings to God". To that end, chemistry was judged more suitable than other more "mature" sciences, such as mechanics or astronomy. Nevertheless, even if the Berkeleian reading of Newton and Homberg was not correct, Peterschmitt shows that there were good historical reasons to think that it was right. That is to say, Berkeley's reading was "the common way of reading" in his time.

Timo Airaksinen and Silvia Parigi both acknowledge the strong influence that Newton's *Opticks* (especially some *queries*) had on *Siris*, and Berkeley's constant approval and admiration towards the Newtonian natural philosophy in its whole. But, while Airaksinen points out a tenseness (or rather a paradox) between Berkeley's critique of "mechanistic science of his own day" and his intention "to make a contribution" to corpuscular physics, Parigi refuses the dominant identification of corpuscularianism with mechanism. In her opinion, neither Newton nor Berkeley were mechanistic philosophers, though both of them defended some version of corpuscular philosophy.

Moreover, Airaksinen stresses Newton's difficulties in explaining both gravity and attraction (i.e. the core of his physics) "in terms of the properties of this occult medium, aether", about which Newton himself had to admit (in *query* 21): "I do not know what it is". Berkeley too, a "Newtonian natural philosopher", was not able to give a satisfying *mechanistic* explanation of gravity in terms of an alleged *corpuscular* light/fire/aether (in the *Appendix* to his essay, Airaksinen offers a useful list of synonyms and expressions used to refer to it). "The problems of interpretation are immense", Airaksinen concludes: "how light/fire actually works as a causal factor in natural philosophy requires a separate study".

But what would happen if the *mechanistic* explanatory model was not the *unique* one, and if Berkeley had not adhered to that paradigm? What if *corpuscularianism* should not simply be considered as a synonym of *mechanism*? If a corpuscular, but non-mechanistic natural philosophy did exist, and it was in fact Newton's and Berkeley's theory? Parigi maintains that it is necessary to distinguish between two different models of explanation: a historically dominant causal paradigm (*scire per causas*), and a minor analogical theory, based on the hermeneutic value of signs (*scire per signa*). While Descartes and Newton – that is to say, the two opposite scientific authorities in Berkeley's time – adhered to the dominant model, Berkeley was deeply attracted by the other one, going back to the ancient Stoics: there is wide textual evidence showing this, from his youthful works to (above all) *Siris*. That is the reason why Berkeley may consistently be, at the same time, a corpuscularian and a critic of the mechanistic, Cartesian conception of the world: there are not, therefore, any drastic changes of mind or turning points in his thought. Its unity and consistency may be historically proved once again. Looking closer at Berkeley's sources (and, perhaps, to Newton's as well), it may happen to find out some forgotten authors, like Marsilio Ficino, a leading figure of Renaissance Neo-Platonism: his (and the Renaissance) widespread concept of "spirit" was constantly present in Berkeley's mind in his "esoteric" speculations about aether.

The five essays collected in the last section also aim at widening the traditional historical background of Berkeley's philosophy, in order to better comprehend it. Daniele Bertini proposes a comparison between Berkeley's philosophy of religion – based on the mere experience of God that necessarily belongs to each spirit, provided that it exists – and Scholastic rational theology. In Berkeley's opinion, the schoolmen made the same mistake as the freethinkers: an excessive attention to details, that prevented them from catching the "universal meaning of any knowledge". Bertini and Airaksinen agree on the little interest Berkeley showed towards Biblical exegesis, but while Airaksinen draws the picture of a man of science, definitely far from the abhorred enthusiasm,[17] Bertini insists on faith, considered as the fundament of Berkeley's religion. In his opinion, theology should be a "way of life", similar to "a kind of wisdom", rather than "a positive science". To sum up: "theology is the experience of the divine".

[17]Airaksinen's polemic is directed versus Patricia Fara's interpretation of Berkeley's philosophy.

David Berman gives a completely different, original portrait of the Bishop: he paints him as "a philosopher of little or no religious faith". Though this assertion may sound paradoxical, Berman supports it with historical and psychological arguments: according to him, in his "heroic years" (1705–1707), Berkeley developed his immaterialist philosophy in order to avoid scepticism, that tempted him in his youth and was felt as a sort of "dark night of the soul". The sceptical attitude was a main outcome of the new science and the new representationalistic or dualistic theories of ideas (starting from Descartes): both of them had created "a gap between what we experience and what exists", but it was only Berkeley who did not feel at ease with that. The character of the "unhappy sceptic", culminating in the Hylas of *Three Dialogues*, would be a proper description of Berkeley himself in his early years: in fact, no ancient or modern author, from Sextus Empiricus to Pierre Bayle, leads us to think that he was unhappy about his (more or less) sceptical philosophy.

Therefore, in Berman's opinion, Berkeley was the first modern philosopher who was painfully aware of "the dire psychological consequences of the sceptical gap of the new science". He was "depressed by the sceptical spectre": the outcome of this unhappy state of mind and unsatisfactory epistemological theory was the *esse est percipi* thesis. That is to say, the "sceptical gap" should not be overcome by faith, as Malebranche and Bayle thought, but exclusively by reason, as Hobbes and Spinoza (two of the "three devils" of the early eighteenth century, along with Machiavelli) maintained. Nevertheless, Berkeley succeeded in finding a way between the "devils" and the "deep blue sea" of faith, that is to say (leaving these suggestive metaphors apart) between fideistic and theological arguments – which he did not like to refer to – and irreligion, heresy or atheism. I think that we should always remember a usually overlooked, but meaningful sentence that Berkeley (under the pseudonym of Misatheus) wrote in a letter published in the *Guardian* on Saturday 21 March 1713: "reason abandons men that would employ it against religion".[18]

Geneviève Brykman highlights the "equivocal" presence of Spinoza in Berkeley's writings (from *Philosophical Commentaries* to *Alciphron* and *Siris*), both as a "similar" thinker – because of his monist doctrine of the infinite substance and his critique of abstract general ideas, that Berkeley shared – and as an adversary – he was undoubtedly considered, in Crito's words, as "the great leader of our modern infidels".

Caterina Menichelli faces a connected topic: the charges of spinozism and atheism levelled towards Berkeley's philosophy at its early reception, from 1718 to 1751, for example by the French Jesuit Tournemine (in 1718), by an anonymous reviewer in *Acta Eruditorum* (1727), and then by some more famous thinkers, as Andrew Baxter, in his *Enquiry into the Nature of the Human Mind* (1733), Andrew Ramsey, in his *Philosophical Principles* (1748), and Robert Clayton, Bishop of Clogher, in his *Essay on Spirit* (1751).

[18] A.A. Luce, Introduction to *Essays in the Guardian,* in George Berkeley, *The Works*, ed. A.A. Luce and T.E. Jessop (London, Edinburgh: Nelson, 1948–1957), vol. 7, 176.

The last essay, by Sébastien Charles, deals with a less analyzed topic: Berkeley's conception of the animal. This was quite an uncomfortable argument for Berkeley, because he had to avoid two equal and opposite "heretical" positions: Descartes' well-known theory of the animal as an automaton, and the free-thinkers' attribution of a soul to the animal, which therefore dangerously ends up being like man himself. It is not surprising that Berkeley dealt with that topic above all in *Philosophical Commentaries*, in the essays published in the *Guardian*, in *Alciphron* and *Siris*, instead of writing about that in his "major" works. Charles solves Berkeley's apparent inconsistencies (the animal perceives, therefore it should have a perceiving mind or soul) by referring to the Neoplatonic concept of the Great Chain of Being: that is to say, Berkeley does not think in terms of Cartesian dualism, but is rather inspired by an "ancient dualism", both Platonic and Aristotelian, which distinguishes rational (immortal) soul and sensitive (mortal) soul, and by Leibniz's theory of the hierarchy of souls. It is not incoherent, thus, to attribute to animals sensation, imagination and even a soul, provided that they are conceived as different from ours *in their degree*.

The coherence of Berkeley's philosophy – which does not exclude, of course, persistent obscurities and elements of tenseness – stands out, if we are able to widen our historiographical perspective, and acknowledge the influence of ancient and Renaissance philosophy on Berkeley's thought.

Part I
Interpretations of Berkeley's Philosophy

Chapter 1
How Berkeley's Works Are Interpreted

Stephen H. Daniel

There are different kinds of studies of Berkeley. Some focus on specific areas of his thought; some provide overviews.[1] Of the overviews, some are arranged according to the chronology of his individual works; others are arranged according to topics.[2] Internal, analytic studies examine the cogency of his arguments and show how different interpretations of his texts handle criticisms raised by recent commentators; historical studies describe the background assumptions that inform his thinking.

More often than not, historical studies propose that we focus on issues and ways of thinking characteristic of seventeenth- and eighteenth-century philosophy in order to appreciate Berkeley's insights. Generally that has meant that Berkeley is interpreted primarily in terms of Locke. A. A. Luce and Harry Bracken, however,

[1] Representative studies of specific areas of Berkeley's thought include: Margaret Atherton, *Berkeley's Revolution in Vision* (Ithaca: Cornell University Press, 1990); George Pappas, *Berkeley's Thought* (Ithaca: Cornell University Press, 2000); Daniel Flage, *Berkeley's Doctrine of Notions* (London: Croom Helm, 1987); Douglas Jesseph, *Berkeley's Philosophy of Mathematics* (Chicago: University of Chicago Press, 1993); C. George Caffentzis, *Exciting the Industry of Mankind: George Berkeley's Philosophy of Money* (Dordrecht: Kluwer Academic Press, 2000); and Paul Olscamp, *The Moral Philosophy of George Berkeley* (The Hague: Martinus Nijhoff, 1970). Overviews include: David Berman, *George Berkeley: Idealism and the Man* (Oxford: Clarendon Press, 1994); A. C. Grayling, *Berkeley: The Central Arguments* (London: Duckworth, 1986), and Geneviève Brykman, *Berkeley: philosophie et apologétique* (2 vols.; Paris: Vrin, 1984).

[2] Examples of chronologically arranged overviews include: John Wild, *George Berkeley: A Study of His Life and Philosophy* (Cambridge, MA: Harvard University Press, 1936); and Dominique Berlioz, *Berkeley: un nominalisme réaliste* (Paris: Vrin, 2000). Topically arranged overviews include: Ian C. Tipton, *Berkeley: The Philosophy of Immaterialism* (London: Methuen, 1974); George Pitcher, *Berkeley* (London: Routledge and Kegan Paul, 1977); and Kenneth P. Winkler, *Berkeley: An Interpretation* (Oxford: Clarendon Press, 1989).

S.H. Daniel (✉)
Department of Philosophy, Texas A&M University, TX 77843-4237, USA
e-mail: sdaniel@philosophy.tamu.edu

S. Parigi (ed.), *George Berkeley: Religion and Science in the Age of Enlightenment*, International Archives of the History of Ideas/Archives internationales d'histoire des idées 201, DOI 10.1007/978-90-481-9243-4_1, © Springer Science+Business Media B.V. 2010 have argued that Malebranche is just as viable a source for Berkeley's thought.[3] Charles McCracken has highlighted Berkeley's similarities with English Malebrancheans (e.g., John Norris).[4] And Richard Popkin and Geneviève Brykman have traced some of Berkeley's ideas to Pierre Bayle.[5] But since Locke does not differ from Malebranche or Bayle regarding some fundamental ontological and epistemological presuppositions, in many historical treatments Locke is still used as the key for understanding Berkeley.

Indeed, for many commentators, the assumption that Berkeley shares doctrines with Locke, Descartes, or Malebranche (e.g., about mental substances) is so central that it has been called the official or standard approach to the study of Berkeley. Proponents of this approach acknowledge that, unlike Locke and Malebranche, Berkeley maintains that only minds and ideas exist; but they usually take it for granted that Berkeley agrees with Locke and Malebranche at least about the fact that minds and ideas are things that exist. As is well known, though, Berkeley insists that spirits and ideas are so different that to say that they are *things* or even that they *exist* fails to appreciate how misleading such beliefs can be (PHK 89, 142).[6] Nonetheless, defenders of the standard approach claim that Berkeley's engagement with the issues raised by Locke or Malebranche indicates that the principles for interpreting Berkeley should be based on Lockean or Malebranchean ways of thinking. Accordingly, they fail to consider how he appeals to the vocabulary of Locke and Malebranche to undercut and supplant their doctrines with his own distinctive views. Those views become evident only when his discussions of Cartesian or Lockean topics (particularly in his *Treatise concerning the Principles of Human Knowledge* and *Dialogues between Hylas and Philonous*) are understood in the context of the entire corpus of his writings. By using the *Principles* and *Dialogues* to authorize Berkeley's "considered" views, the standard approach minimizes how his unpublished remarks and other publications raise doubts about the propriety of interpreting his central insights in Cartesian, Malebranchean, or Lockean terms.

The standard reading of Berkeley thus proposes that we ignore how his doctrines of mind and ideas must be understood within his entire corpus. It invokes a strategy

---

[3] A. A. Luce, *Berkeley and Malebranche: A Study in the Origins of Berkeley's Thought* (London: Oxford University Press, 1934); A. A. Luce, *The Dialectic of Immaterialism* (London: Hodder and Stoughton, 1963); and Harry M. Bracken, *Berkeley* (New York: St. Martin's Press, 1974), 16–18.

[4] Charles J. McCracken, *Malebranche and British Philosophy* (Oxford: Clarendon Press, 1983).

[5] See Richard Popkin, "Berkeley and Pyrrhonism," *Review of Metaphysics* 5 (1951): 223–246; and Geneviève Brykman, "Berkeley: sa lecture de Malebranche à travers le *Dictionaire* de Bayle," *Revue internationale de philosophie* 114 (1975): 496–514.

[6] Citations from *The Works of George Berkeley* [*Works*], ed. A. A. Luce and T. E. Jessop (9 vols.; London: Thomas Nelson, 1948–1957) include *A Treatise concerning the Principles of Human Knowledge* [PHK section] and *Dialogues between Hylas and Philonous* [DHP page], vol. 2; *Alciphron* [Alc dialogue and section], vol. 3; and *Siris* [section], vol. 5; *Passive Obedience* [PO section], vol. 6. References to Berkeley's *Notebooks* [NB entry], and *De Motu* (Luce translation) [DM entry], and *Theory of Vision Vindicated and Explained* [TVV section] are from his *Philosophical Works*, ed. Michael. R. Ayers (Rutland, VT: Charles E. Tuttle, 1992).

of interpretation that is unacknowledged and unchallenged – even when it ends up portraying Berkeley as confused, contradictory, or prone to fundamental changes in his positions.[7] In the standard interpretation, the focus on Berkeley's *Principles* and *Dialogues* justifies the appeal to Descartes, Malebranche, and Locke as the interpretive filters through which his other works are understood, because the *Principles* and *Dialogues* are the works in which he most directly addresses their issues. Interpretations that do not rely on a Cartesian or Lockean framework are usually dismissed as tendentious, far-fetched, or inconsistent with what the text "actually" says.[8] The attempt to avoid an ahistorical reading is thus replaced by the equally pernicious strategy of thinking that Berkeley's citations of Descartes, Malebranche, and Locke indicates that he thinks in their terms as well. When Berkeley says something that is inconsistent with a Cartesian or Lockean position – such as, "the very existence of ideas constitutes the soul" (NB 577) – his comments are then usually interpreted (in the standard approach) to be positions he ultimately rejects – in some instances, just a few pages or days later.[9]

John Roberts has recently thematized such an interpretive strategy, recommending that we elucidate Berkeley's views "by locating them with respect to two traditions of the early modern period that inform and compete with his, those of Descartes and Locke."[10] When we do this, he concludes, we have a much clearer view of the basic features of Berkeleian thought. To make sure that those features are protected from any unnecessary complications in comparing remarks in Berkeley's unpublished *Notebooks* and his published works, Roberts says that we should adopt what he calls "Constraint 1": "When there is a conflict, one should reject early views that the author chose not to publish in favor of later views that the author chose to publish repeatedly" (*Mob* 7). Of course, this constraint is intended to be used especially in those instances in the *Notebooks* where Berkeley says something that sounds uncharacteristic of his more well-known views. In those cases, it is obviously easier to explain away the remarks by considering them as ill-conceived and subsequently rejected views, rather than to show how they can be interpreted in a way that is consistent with Berkeley's other comments when they are understood apart from a non-Cartesian or non-Lockean context.

---

[7] For example, see Phillip D. Cummins, "Berkeley on Minds and Agency," in *The Cambridge Companion to Berkeley*, ed. Kenneth P. Winkler (New York: Cambridge University Press, 2005), 209–218.

[8] For example, see Michael R. Ayers, "Berkeley, Ideas, and Idealism," in *Reexamining Berkeley's Philosophy*, ed. Stephen H. Daniel (Toronto: University of Toronto Press, 2007): 27–28; and Marc Hight and Walter Ott, "The New Berkeley," *Canadian Journal of Philosophy* 34 (2004): 8–18.

[9] Cf. Charles J. McCracken, "Berkeley's Cartesian Concept of Mind: The Return through Malebranche and Locke to Descartes," *The Monist* 71 (1988): 597–600; idem, "Berkeley's Notion of Spirit," *History of European Ideas* 7 (1986): 597–602; and Bertil Belfrage, "Berkeley's Four Concepts of the Soul," in *Reexamining Berkeley's Philosophy*, ed. Stephen H. Daniel (Toronto: University of Toronto Press, 2007), 172–187.

[10] John R. Roberts, *A Metaphysics for the Mob: The Philosophy of George Berkeley* (New York: Oxford University Press, 2007), 8. Hereafter: *Mob*.

No doubt, this easy way of reading Berkeley has kept him in the forefront of thinkers to be reckoned with in early modern philosophy. But it has also meant having to tweak the standard presentation of some of his stranger doctrines (e.g., the language of nature, spiritual substances, cosmic fire) to bring them more in line with Descartes or Locke. These efforts have hardly been met with enthusiasm, and in some cases they have even been characterized as annoying. So to avoid a direct challenge to the standard interpretation, some scholars focus instead on Berkeley's *Alciphron, Analyst, Querist,* or *Siris.*[11] Because such shifts in focus generally do not affect our understanding of Berkeley's relation to the Cartesian or Lockean project, they are hardly considered a threat to the standard view.

But what if there were a non-Cartesian, non-Lockean way to understand Berkeley's writings without having to assume that he changes his fundamental insights or at times simply contradicts himself? That is, suppose that Berkeley says exactly what he means, and that all of his claims are consistent with one another. In that case, we could say that there are occasions when he chooses to modify his expressions because he recognizes how his views are misunderstood by those with whom he disagrees. Such choices, though, would not indicate changes in his views; rather, they would reveal only a willingness on his part to accommodate the limited perspectives of some of his contemporaries.

This way of reading Berkeley would, of course, require that we maintain the same scholarly standards used in any interpretation of his work. But more importantly, it would make explicit its commitment to the principle of charity. So if reading Berkeley in the context of a Cartesian or Lockean way of thinking leads us to conclude that he is confused or needs to change his views or fails to explain why he would subsequently be attracted to seemingly unrelated doctrines, then we should look for another way to interpret his thought. Such a practice would, no doubt, challenge the standard view and open up a number of strategies for interpreting Berkeley by considering his doctrines in the context of contemporaries with whom he is seldom linked.[12]

[11] See *Alciphron, or the Minute Philosopher: In Focus,* ed. David Berman (NewYork: Routledge, 1993; Roomet Jakapi, "Emotive Meaning and Christian Mysteries in Berkeley's *Alciphron,*" *British Journal for the History of Philosophy* 10 (2002): 401–411; Olscamp, *Moral Philosophy of Berkeley*; Jesseph, *Berkeley's Philosophy of Mathematics*; Caffentzis, *Berkeley's Philosophy of Money*; Timo Airaksinen, "The Path of Fire: The Meaning and Interpretation of Berkeley's *Siris,*" in *New Interpretations of Berkeley's Thought,* ed. Stephen H. Daniel (Amherst, NY: Humanity Books, 2007), 261–281; and Costica Bradatan, *The Other Bishop Berkeley: An Exercise in Reenchantment* (New York: Fordham University Press, 2006).

[12] I have argued elsewhere that Berkeley's doctrines can be fruitfully interpreted by comparing them to the Stoics, the seventeenth-century followers of the Renaissance logician Peter Ramus, Jonathan Edwards, Francisco Suárez, Baruch Spinoza, and G. W. Leibniz – all of whom develop philosophies that are markedly different from those of Descartes, Malebranche, and Locke. See Stephen H. Daniel, "Stoicism in Berkeley's Philosophy," in *Berkeley's Lasting Legacy: 300 Years Later,* eds. Bertil Belfrage and Timo Airaksinen (Newcastle upon Tyne: Cambridge Scholars, forthcoming); "The Ramist Context of Berkeley's Philosophy," *British Journal of the History of Philosophy* 9 (2001): 487–505; "Edwards, Berkeley, and Ramist Logic," *Idealistic Studies* 31

Informing a shift away from the traditional or received strategy of interpreting Berkeley are the following historiographic assumptions:

- First, the fact that Berkeley's published comments appear to be inconsistent with one another does not mean that they *are* inconsistent. It is arbitrary and even arrogant to assume that it is more likely that he is confused or inconsistent than that we have failed to understand the nuances of his position.
- Second, the fact that Berkeley's published comments appear to be inconsistent with some of his unpublished comments does not permit us to conclude that the unpublished comments represent views that he rejects or doubts. It is obviously easier to dismiss his unpublished remarks rather than to do the hard work of discerning how seeming inconsistencies can be overcome. Besides, authors are not required to publish all of their ideas, especially when they suspect (as Berkeley quickly discovered) that readers locked into a Cartesian or Lockean mindset will misinterpret them. When commentators refuse to treat Berkeley's private 1708 notes on a par with his "considered" opinions published a year or two later, or cite an unpublished remark only when it supports a favored interpretation, or ignore *De Motu* (1721), *Alciphron* (1732), and *Siris* (1744) in favor of the "mature" works of 1709–1713 – they reinforce the bias in favor of thinking that Berkeley's importance relies on reading him in a Cartesian or Lockean context.
- Third – and this one should be apparent to anyone familiar with the fallacy of the appeal to authority – the fact that an interpretation has become the official or received view does not guarantee its correctness – especially if it is based on violations of the two prior principles.

My purpose in raising these points is to emphasize how Berkeley's published works in philosophy, religion, mathematics, science, economics, and politics are often understood apart from one another. Considering how his unpublished writings are relegated to a secondary status, and how his "early" (pre-1709) and "late" (post-1737) writings are treated as unimportant for grasping his central insights, and how he is supposed to have changed his mind even in his most well-known works, it is no wonder that commentators complain about not being able to determine Berkeley's *real* position on certain issues.

All of this can be avoided, however, if we treat Berkeley's writings as consistent with one another, and instead of thinking that he rejects his own ideas, we think that he merely refocuses his attention and adapts his ways of speaking to accommodate certain contexts or to clarify earlier remarks. His *Notebooks* provide the best opportunity for testing such a strategy. However, because Berkeley's writings are so

(2001): 55–72; "Berkeley, Suárez, and the *Esse-Existere* Distinction," *American Catholic Philosophical Quarterly* 74 (2000): 621–636; "Berkeley and Spinoza," *Revue philosophique de la France et de l'étranger* 135 (2010): 123–134; and "The Harmony of the Leibniz-Berkeley Juxtaposition," in *Leibniz and the English-Speaking World*, ed. Stuart Brown and Pauline Phemister (Dordrecht: Springer, 2007): 163–180.

varied, this way of interpreting him also invites us to consider a wealth of previously unexplored connections between topics such as mind, political obligation, divine grace, and money. And by failing to assume the overall integrity of Berkeley's thought, we miss out on the opportunity to understand his immaterialist and idealist views in ways that are not fraught with confusion, changes of heart, or contradictions.

To test my proposal, I appeal to three examples. The first concerns Berkeley's supposed change regarding his view that "Bodies taken for Powers do exist when not perceived" (NB 293a; also NB 52), in that bodies can be understood as collections of powers to cause our thoughts (NB 80, 282). Some interpreters (e.g., A. A. Luce, Charles McCracken, A. C. Grayling, Robert McKim) claim that he later drops his belief in powers when he decides "not to mention the Combinations of Powers but to say the things, the effects themselves, to really exist even when not actually perceived" (NB 802).[13] However, instead of dismissing Berkeley's initial insight as an unfortunate gaff, we can just as easily say that he realizes that his attempt to reframe talk of bodies in terms of the source of regular patterns of experience is destined to be misunderstood by Cartesian and Lockean readers because of their inability or unwillingness to think of powers in terms other than as qualities inhering in material substances.

Berkeley's doctrine of powers is therefore not something he rejects or changes. Indeed, in the *Dialogues* Philonous allows for talk of powers: "I assert as well as you that, since we are affected from without, we must allow Powers to be without, in a Being distinct from ourselves" (DHP 240; also 242, NB 41, TVV 11–12); and in *De Motu* he notes that the real powers of bodies reside in their cause (DM 71). As Berkeley maintains early on in his *Notebooks* and consistently throughout his career, that cause or "active power" (NB 131, 155) is nothing other than the will of "one simple perfect power" (NB 282). His decision "not to mention" powers thus represents no change, only an accommodation to those who otherwise do not understand how bodies (the proper domain of natural philosophy) cannot be described by appealing to metaphysical principles.[14]

The second example concerns Berkeley's much discussed "bundle theory" of mind. Some readers have interpreted his remark that the "mind is a congeries of perceptions" (NB 580) as an anticipation of Hume's theory that the mind is a bundle of ideas. But such a view seems to contradict his published pronouncements that the mind is a spiritual substance. To resolve the conflict, commentators have suggested three ways to handle Berkeley's so-called two concepts of mind.

[13] Cf. Luce, *Dialectic of Immaterialism*, 134–135, 140, 154; Charles J. McCracken, "What *Does* Berkeley's God See in the Quad?" *Archiv für Geschichte der Philosophie* 61 (1979): 284–285; Grayling, *Berkeley*, 97–98, 101; and Robert McKim, "Berkeley's Notebooks," in *The Cambridge Companion to Berkeley*, ed. Kenneth P. Winkler (New York: Cambridge University Press, 2005), 87–88.

[14] See Stephen H. Daniel, "Postface: les limites de la philosophie naturelle de Berkeley," in *Science et épistémologie selon Berkeley*, ed. Sébastien Charles (Québec: Presses de l'Université Laval, 2004), 165–170.

According to the first approach, Berkeley is said to have always endorsed a Cartesian notion of the self as a spiritual substance, and his mention of the bundle theory is simply an indication of passing musings about a theory he ultimately rejects.[15] Proponents of the second approach argue that the bundle theory was Berkeley's real position, but for prudential reasons he gave lip service to the view that the mind is a Cartesian or Lockean substance.[16] Those who adopt the third approach say that Berkeley accepted the bundle theory for a short time and then rejected it in favor of his published view.[17] Despite their differences, all three strategies grant that the Cartesian and Humean concepts of mind are incompatible, and that is what drives the effort to show how Berkeley could not have proposed both.

Of course, the major flaw in these strategies is that they impose a Cartesian or Humean way of thinking about spiritual substance or mind onto Berkeley, and they refuse to acknowledge how he could develop a distinctive doctrine that does not draw on either view and does not attempt to reconcile them. These strategies of interpretation thus fail to appreciate how Berkeley objects fundamentally to describing the mind or spiritual substance as a *thing* that thinks or wills – not only because such a description is unnecessary but also because it easily misleads us into thinking that mind can be an object of thought (i.e., an idea) or can even be said to be a thing that thinks or wills:

> Say you the mind is not the perceptions but that thing which perceives. I answer, you are abused by the words *that* and *thing*: these are vague, empty words without a meaning .... Say you there must be a thinking substance, something unknown which perceives and supports and ties together the ideas. Say I, make it appear there is any need of it, and you shall have it for me .... If you ask what thing it is that wills, I answer if you mean idea by the word *thing* or anything like an idea, then I say tis no thing at all that wills. This how extravagant soever it may seem yet is a certain truth. We are cheated by these general terms, *thing*, *is*, etc. Again, if by *is* you mean is perceived or does perceive, I say no thing which is perceived or does perceive wills .... Substance of a spirit is that it acts, causes, wills, operates, or if you please (to avoid the quibble that may be made on the word *it*), to act, cause, will, operate; its substance is not knowable, not being an idea. (NB 581, 637, 658–59, 829)

[15] See Luce, *Dialectic of Immaterialism*, 24–33, 173; and Marc Hight and Walter Ott, "The New Berkeley," *Canadian Journal of Philosophy* 34 (2004): 8–11. This strategy relies on the now discredited assumption that Berkeley's marginal "+" mark indicates a position he comes to reject.

[16] See Colin M. Turbayne, "Berkeley's Two Concepts of Mind," in *Berkeley: Principles of Human Knowledge, Text and Critical Essays*, ed. Colin M. Turbayne (Indianapolis: Bobbs-Merrill, 1970), 145–160; and Robert G. Muehlmann, "The Substance of Berkeley's Philosophy," in *Berkeley's Metaphysics: Structural, Interpretive, and Critical Essays*, ed. Robert G. Muehlmann (University Park: Pennsylvania State University Press, 1995), 89–105.

[17] See Pitcher, *Berkeley*, 183–189; Charles McCracken, "Berkeley's Notion of Spirit," *History of European Ideas* 7 (1986): 597–602; idem, "Berkeley's Cartesian Concept of Mind," *The Monist* 71 (1988): 597–611; Berman, *Berkeley*, 69–70; Talia Mae Bettcher, *Berkeley's Philosophy of Spirit* (London: Continuum, 2007), 2; Bertil Belfrage, "A New Approach to Berkeley's *Philosophical Notebooks*," in *Essays on the Philosophy of George Berkeley*, ed. Ernest Sosa (Dordrecht: D. Reidel, 1987), 217–230; and idem, "Four Concepts," 174–182.

To think that Berkeley appeals to a description of mind or spiritual substance similar to that of Descartes, Malebranche, Locke, or Hume is merely to adopt a historiographic prejudice that precludes the possibility that Berkeley can invoke the terms they use in his own unique, even "extravagant" ways. No doubt, respecting Berkeley's own use of such terms threatens to make his doctrine of mind less easily compared to those of his contemporaries, but it opens up new ways to understand how that doctrine complements other aspects of his philosophy.

Instead of presuming a Humean context – in which the mind as "a congeries of perceptions" is understood as a mere bundle of *objects* of perception – we can just as easily think that Berkeley is referring to the unity of *acts* of perceiving. After all, he uses the word "perception" as either object or act on numerous occasions (e.g., PHK 5; DHP 195–196). We certainly do not have to conclude that at some point in 1707 or 1708 he briefly held a Hume-like view that he quickly came to see was flawed. To be sure, he subsequently avoids describing mind in those terms, but this says less about his position than it does about his concern that calling mind a congeries will be misunderstood by those who think of mind as an object rather than as the will that objects be identified and related. As in his appeal to "powers," Berkeley here again signals his reluctance to cede the description of mind to those who would interpret him solely in a Cartesian or Lockean context.

When we read, then, that "the very existence of ideas constitutes the soul" (NB 577), we need not think that this contradicts Berkeley's fundamental belief that spirits are radically different from ideas. Instead, it indicates how the very existence of the soul consists in the will that there be the differentiation and association of ideas. As Berkeley observes, "The spirit, the active thing, that which is soul and God, is the will alone .... Tis one will, one act distinguished by the effects. This will, this act is the spirit, operative principle, soul, etc." (NB 712, 788). Spirit is not some thing or object that has a will or that engages in willful activity, for to think of it that way – that is, "by way of idea" (PHK 142) – would reinstate the very substance-mode ontology that Berkeley wants to overthrow. Rather, mind *is* the will, and it does not exist apart from its activities (PHK 98).

In this sense, God's will that there be a distinctive sequence of acts of differentiating and associating ideas in perceptions constitutes the existence of a soul or mind.[18] Accordingly, Berkeley can admit that "there are innate Ideas i.e. Ideas created with us" (NB 649), for God's creation of minds is the very same act as his specification of their ideas. As Berkeley later notes regarding Parmenides and Plato, "To understand and to be ... are the same thing .... Whence it follows that mind, knowledge, and notions, either in habit or in act, always go together" (*Siris* 309). It is this interrelation of minds and their ideas that is captured in Berkeley's doctrine of innate ideas.

[18] See Stephen H. Daniel, "Berkeley's Semantic Treatment of Representation," *History of Philosophy Quarterly* 25 (2008): 41–55; and idem, "Berkeley's Stoic Notion of Mind," *New Interpretations of Berkeley's Thought*, ed. Stephen H. Daniel (Amherst, NY: Humanity Books, 2007), 203–230.

To say that Berkeley endorses a doctrine of innate ideas – especially when he is generally assumed to follow Locke in rejecting them – is hardly a commonplace in Berkeley scholarship.[19] Indeed, when confronted with Berkeley's claim that we are created with innate ideas, commentators typically dismiss NB 649 as an aberration.[20] But as with powers and the bundle theory, Berkeley does not change his mind on innate ideas. Rather, he adapts its salient features to his idealism, redefining innatism in a way that permits him to argue that different minds perceive the "same" ideas and experience the same moral sentiments – not because the objects of their perceptions (i.e., the standard Cartesian or Lockean innate ideas) are identical, but because God wills that their acts of perceiving are in concert with one another (PO 25; Alc I.14).[21] In this way, the issue of innate ideas serves as a third example of how an otherwise puzzling doctrine in Berkeley's philosophy can be explained without portraying him as confused or dismissing his texts as unrepresentative of his "considered" views.

Admittedly, Berkeley's way of thinking about the mind's relation to its ideas is different from that found in Descartes, Malebranche, or Locke, so it is hardly surprising to see how few commentators have appreciated its significance for understanding the overall cohesiveness of his thought. Geneviève Brykman, however, points out that Berkeley's description of nature as a language links our ideas essentially to mind by embedding them in an always already discursive context.[22] This linguistic characterization of mind challenges the standard Cartesian or Lockean strategies for interpreting Berkeley by refusing to assume that minds and ideas are ontological givens. Instead, she sees Berkeley as developing a philosophy in which things in the world are identified by being *differentiated* in the "veil" of the divine language that minds enact.

No doubt, this strategy for interpreting Berkeley is unfamiliar to some historians of modern philosophy because it requires thinking in the Stoic terms of difference and propositional expression rather than the Platonic/Aristotelian terms of identity and predication on which Cartesian and Lockean strategies draw.[23] That might

---

[19] But see Louis Loeb, *From Descartes to Hume: Continental Metaphysics and the Development of Modern Philosophy* (Ithaca: Cornell University Press, 1981), 69–70; and Daniel Flage, "Berkeley, Individuation, and Physical Objects," in *Individuation and Identity in Early Modern Philosophy: Descartes to Kant*, ed. Kenneth F. Barber and Jorge J. E. Gracia (Albany: SUNY Press, 1994), 142.

[20] See Michael R. Ayers, "Was Berkeley and Empiricist or a Rationalist?" in *The Cambridge Companion to Berkeley*, ed. Kenneth P. Winkler (New York: Cambridge University Press, 2005), 48; and Wayne Waxman, *Kant and the Empiricists* (New York: Oxford University Press, 2005), 267n.

[21] See Berkeley, Sermon 10, "On the Will of God," *Works* 7: 130. Cf. Bracken, *Berkeley*, 115; and Flage, *Doctrine of Notions*, 188.

[22] See Geneviève Brykman, *Berkeley et le voile des mots* (Paris: Vrin, 1993), 52, 70–74, 422–423; and Henri Bergson, "Philosophical Intuition," in *The Creative Mind* [CM], trans. Mabelle L. Andison (New York: Philosophical Library, 1946), 140–141.

[23] Cf. Gilles Deleuze, *The Logic of Sense* [1969], trans. Mark Lester and Charles Stivale; ed. Constantin V. Boundas (New York: Columbia University Press, 1990), 7, 105.

explain why many readers fail to appreciate how Berkeley's shift in focus to the language of nature requires that we think of things first and foremost as signs that depend on being perceived specifically as *this* or *that* thing – that is, as related to other things.

This shift in perspective – from simply assuming that things have their identity apart from their role in a communicative network to thinking of things in terms of that network – is at the heart of not only Berkeley's philosophy but also his self-acknowledged Irish identity. For even at 8 years old, he says, he distrusted certain doctrines of English and continental contemporaries (NB 266), seemingly because they are so inconsistent with *common* (i.e., communal) sense. That Irish sensibility is not due to Berkeley's Irish birth. Rather, it refers to his belief that everything has meaning in terms of distinctions and relations that are literally sensible (i.e., as both meaningful and based on sense experience).

This commitment to the rhetorical heritage of experience runs throughout Berkeley's work, and it makes his thought stand out from the pronouncements of contemporaries who overlook the practical, sensible, or communicative legacy of things. "We Irishmen," he observes, know that being aware of anything – whether it be a material object or an infinitesimally small point – means invoking a discursive heritage apart from which no identity or reasoning is possible (NB 392–94, 398). As Berkeley sees it, any truly learned person knows that he or she is immersed in and identified by such a heritage. Only a "vulgar" person would think that things are simply *there* in the world, wearing their meaning or significance on their sleeves, indifferently available to all who would gaze upon them. Perhaps the English or the French might be arrogant enough to presume that their linguistic and cognitive heritage captures the universal experience of reality, but the Irish cannot pretend to such vulgarity. That is why when Berkeley says that we should speak with the vulgar and think with the learned, he means that even if we adopt the language of a Malebranche or Locke, we can still subvert it by drawing attention to its contingency.

Underlying Berkeley's philosophy (including his epistemology, metaphysics, philosophy of science, and ethics), then, is his theory of meaning. In that theory, statements are intelligible in virtue of how they function within a system of signs. Mind is central in this account, because it is the activity of differentiating and relating things in the world as terms in the language of nature. Minds are not so much things addressed in the language (though they can be described derivatively in such terms) as the principles that identify how things in the world are intelligible. Mind is thus the signification, the intelligibility, or more specifically, the linguisticality of the world.

Hence, it is in terms of mind that all of Berkeley's works – including those sometimes considered on the periphery of his philosophy, such as *Passive Obedience*, *Alciphron*, *The Analyst*, *The Querist*, and *Siris* – take on greater importance, in that they reveal how freedom, morality, value, and beauty are explicable in the same terms that characterize Berkeley's epistemology and metaphysics. For as he notes early on, "there can be no perception, no idea without will, being there are no ideas so indifferent but one had rather have them than annihilation, or annihilation than them . . . there being no ideas perfectly void of all pain and uneasiness" (NB 833). All perception is thus unavoidably affective, because everything we experience – whether or not we recognize it as such – is significant in virtue of its place in a system of signs. By embedding the discussion of mind in the context of a system of signs, Berkeley (unlike his contemporaries) portrays mind as the will or intention that things be identified by being differentiated and related rather than as something distinct from already differentiated things.

In short, the way to interpret Berkeley's works is (as he recommends to Samuel Johnson) to read them in order.[24] In his *Notebooks* he sketches out the issues that will occupy his subsequent works, including examinations of: (1) how things are intelligible within networks of sign relations (*New Theory of Vision*, 1709); (2) what it means for things to exist (*Principles* I, 1710); (3) what minds are and how they relate to one another (*Principles* II, lost; *Passive Obedience*, 1712); and (4) how God's act of creation is related to ideas and finite minds (*Dialogues*, 1713). Later works describe God's ordering of events in terms of how objects are regulated by the laws of nature (*De Motu*, 1721) and how the minds that experience those objects are morally responsible and still free (*Alciphron*, 1732). The *Theory of Vision Vindicated* (1733) highlights the different ways in which finite minds confront the challenge of ordering experience (i.e., suggestion, inference); and the second edition additions to *Principles* and *Dialogues* (1734) reveal how the inherently semiotic or intentional character of mind is captured in the vocabulary of notions. In his works on mathematics and money – *The Analyst* (1734), *A Defence of Free-thinking in Mathematics* (1735), and *The Querist* (1735–1737) – Berkeley indicates how "notions" provide an intermediate concept between the signifying relations of ideas as determined by the divinely-instituted language of nature and those established through merely human convention.[25] Indeed, notional signification is Berkeley's accommodation to our fallen human situation, in that the notions on which we rely in mathematics (e.g., proportion, infinity) and economics (e.g., money) are neither purely natural nor simply conventional, neither divine nor grounded merely in extra-systemic reference. In *Siris* (1744) this appeal to an intermediate way of speaking is extended to the minute corpuscles responsible for the efficacy of tar water. As the *ultimate* cause of all things, however, mind is the principle by which things are differentiated from and related to one another. Because our fallen nature precludes our thinking simply in those terms, Berkeley appeals to "pure elementary fire" (or "acidity") as the *instrumental* cause of such differentiation.

In *Siris*, Berkeley's philosophy comes full circle back to its original enquiry into how existence has meaning, in that it characterizes the central activity of mind not as an impulse toward homogenous unity but as the effort to achieve harmony in an

---

[24] Berkeley to Johnson, 30 March 1730, in *Philosophical Works*, 355.

[25] See C. George Caffentzis, "Algebraic Money: Berkeley's Philosophy of Mathematics and Money," *Berkeley Studies* 18 (2007): 3–23.

ever-increasing variety of expressions. Instead of removing differences, mind and fire create multiplicity – but always within patterns of regularity. Just as in *Alciphron* (IV.12), mind in *Siris* is recognized in the copious, articulate, and varied expressions of will. In *Siris*, as in Berkeley's other works, mind is portrayed as the principle of intelligibility and meaning precisely because it is the principle of differentiation whereby the multiplicity of things in the world is expressed in and as a language.

In Berkeley's last major work, mind, language, and fire thus come together in a way consistent with his early insistence on heterogeneity among the senses. As *Siris* makes clear, God himself is differentiated in the Trinity, and it is this feature of mind's inherent otherness that makes possible the effort to will that there be light – not only in terms of understanding but also in the fiery creation of beings. That theme is there at the beginning of Berkeley's philosophical career, and it is also there at the end. By thinking of it as a central organizing motif, we can appreciate more clearly the unity of his thought and resist the temptation to imagine contradictions where there are none.

Chapter 2
Berkeley's Metaphysical Instrumentalism[1]

Marc A. Hight

Berkeley is widely held to be a scientific instrumentalist, but the scope of his instrumentalism has been repeatedly brought into question. Some have asserted that Berkeley capitulated wholesale to a form of external realism at the end of his life,[2] others have supposed principled reasons for thinking that Berkeley is an instrumentalist about some things and not about others. Lisa Downing, for instance, has argued that Berkeley is an instrumentalist about forces but not about corpuscles,[3] and Douglas Jesseph contends that Berkeley rejects mathematical instrumentalism despite being a stronger instrumentalist in the sciences.[4] Here I suggest that there is a principled form of instrumentalism one may reasonably attribute to Berkeley, such that one need not claim that Berkeley is sometimes an instrumentalist and sometimes not; there is a consistent position that explains the variety of moves Berkeley makes.

2.1 Enter Instrumentalism

First we require a clear sense of what we mean by 'instrumentalism' to preclude our discussion from devolving into verbal games. Here I start by following the able characterization provided by Jesseph, who starts with the basic view. "Instrumentalism,

---

[1] Work on this paper was made possible thanks to a 2007–2008 Fulbright Fellowship. Thanks also to Roomet Jakapi, Luc Peterschmitt , and an anonymous referee for some insightful suggestions that have improved the paper.

[2] Catherine Wilson, "Berkeley and the Microworld," *Archiv für Geschichte der Philosophie* 76 (1994): 37–64.

[3] Lisa Downing, "*Siris* and the Scope of Berkeley's Instrumentalism," *British Journal for the History of Philosophy* 3 (1995): 279–300.

[4] Douglas Jesseph, *Berkeley's Philosophy of Mathematics* (Chicago: University of Chicago Press, 1993): 213.

M.A. Hight (✉)
Department of Philosophy, Hampden-Sydney College, Hampden-Sydney, VA 23943, USA
e-mail: mhight@hsc.edu

S. Parigi (ed.), *George Berkeley: Religion and Science in the Age of Enlightenment*,
International Archives of the History of Ideas/Archives internationales d'histoire des idées 201,
DOI 10.1007/978-90-481-9243-4_2,

broadly speaking, is the doctrine that a theory can be accepted and applied for reasons of utility, even if the claims made in the theory or its application are not accepted as literally true."[5] One immediate concern with this definition when applied to Berkeley is to fix what one means by 'literally true.' With an immaterialist system, truths concern ideas and their orderings. To say that bodies are attracted towards one another according to an inverse square law is true, provided that one understands the proposition to involve regularities in what we experience in the perception of sensory ideas *and nothing further*. There are no deeper causal explanations than the regularities found in our sensory experiences. Materialists make the mistake of wanting a kind of realism that appeals to a causal ordering beyond what we perceive. As a result, they posit the existence of occult things like 'forces' to undergird our ideas. Berkeley remarks in the *Notebooks*, "The supposition that things are distinct from Ideas takes away all real Truth, and consequently brings in a Universal Scepticism, since all our knowledge and contemplation is confin'd barely to our own Ideas."[6] The order of ideas – the patterns we find in experience – are real and we do not *need* to appeal to a 'deeper' level to gain scientific knowledge.

Truth in Berkeley's system concerns a correspondence to God's decreed ordering of sensory ideas. If I make a claim about the world, then either my claim corresponds to what God has willed in the ordering of ideas (and my claim is true), or it does not. Truth for Berkeley concerns the order of ideas. He reveals his view in certain passages and the following is typical.

> But, say you, it sounds very harsh to say we eat and drink ideas, and are clothed with ideas. I acknowledge it does so, the word *idea* not being used in common discourse to signify the several combinations of sensible qualities, which are called *things*; and it is certain that any expression which varies from the familiar use of language, will seem harsh and ridiculous. But this doth not concern the truth of the proposition, which in other words is not more than to say, we are fed and clothed with those things which we perceive immediately by our senses.[7]

In this passage Berkeley is not reflectively outlining his theory of truth, but here and elsewhere it is apparent that what it means for a proposition to be true concerns the accuracy with which that proposition depicts ideas and their various relations to one another.[8] For Berkeley to be an instrumentalist in the basic sense implies that science accepts theories for their utility without referring to anything other than the regularities we find in our perception of sensory ideas.

[5] Jesseph, *Berkeley's Philosophy of Mathematics*, 76.

[6] PC 606. All citations from Berkeley are from *The Works of George Berkeley, Bishop of Cloyne*, eds. A.A. Luce and T.E. Jessop, 9 vols. (London: Thomas Nelson and Sons, 1948–1957). The following abbreviations will be used for convenience: AN: *The Analyst*, 3D: *Three Dialogues Between Hylas and Philonous*, ALC, *Alciphron: or the Minute Philosopher*, PC: *Philosophical Commentaries* (the notebooks), PHK: *Principles of Human Knowledge*, IPHK: *Introduction to the Principles of Human Knowledge*, DM: *De Motu*, and S: *Siris*. Other texts of Berkeley, not abbreviated, are also from this source. Section numbers will be used for the *Principles, De Motu*, and *Siris*; all others will be page numbers from the *Works*.

[7] PHK 38.

[8] There are, of course, complications with propositions that concern active things and notions. Some of those concerns will be engaged later.

Jesseph next moves on to discuss a weaker version of instrumentalism where "a certain body of theory is regarded as false but is nevertheless used for purposes of simplicity and economy."[9] Euclidean geometry is taught to students around the globe, even though we no longer believe it accurately models the world. For Berkeley such a view would be to admit theories that predict future experiences generally well, but not perfectly. Owing to the complicated nature of the world, it might sometimes be best to employ theories that are less complicated because they are 'good enough' for their intended uses. It makes sense to employ Euclidean principles when playing billiards for instance; the added precision one acquires from using more sophisticated and accurate theories makes essentially no difference to even the expert player.

Lastly Jesseph describes what he calls 'an even weaker' version of instrumentalism, which applies only to background assumptions and not theories. According to this view, background variables are treated as irrelevant or insignificant even if the governing theory is thought to be literally true. Jesseph asserts that Berkeley is an instrumentalist of this sort when it comes to geometry. That is, he believes that Berkeley thinks that geometry is true, but not a completely accurate description of what we perceive.[10] These various divisions enable Jesseph to make distinctions within Berkeley's writings and separate his ontological commitments in mathematics and the sciences, although most of the purposes to which Jesseph puts these distinctions are not of concern to us in the present endeavor.

Here I make a case for the claim that Berkeley is an instrumentalist in the basic sense, even about geometry and mathematics generally. As a result, he is an instrumentalist provided he believes that scientific theories should be accepted on the basis of whether they are useful in terms of predicting future experiences. When I claim that Berkeley is a basic instrumentalist, I do not intend that the patterns of experience are not real, but only that there is no *other* underlying cause of those patterns (except God). In particular, we do not need corpuscles or material objects to explain the regular order of ideas we perceive.

2.2 Berkeley's Instrumentalism

We are now prepared to examine the details of my suggested interpretation of his instrumentalism. For Berkeley, both math and science depend on the manipulation of signs that stand for sensory content.[11] And it is a gross error to confuse the usefulness of a theory for its truth, even if they often coincide.[12] Science is one *method* for

---

[9] Jesseph, *Berkeley's Philosophy of Mathematics*, 76.

[10] Jesseph, *Berkeley's Philosophy of Mathematics*, 76–77.

[11] Although it does not depend *solely* on signs, since memory and imagination can play important roles as well. See PC 883.

[12] AN 10, *Works* IV, 70–71.

arriving at the truth, but should not be confused *with* the truth.[13] For example, the concept of force is permissible in science, provided one understands force as a sign for certain kinds of empirical regularities – but there are no metaphysical *things* 'forces.' We can form no idea of force, but the *word* can serve the useful purpose of organizing the experiences we do have.[14] Note that there is nothing about instrumentalism that precludes a theory from being true in its descriptions of the world; it must only be the case that even if its descriptions are true, only the utility of the theory matters.[15] In our vulgar utterances we seem to refer to a material world. Yet all that we actually require to *explain* and *function* in the world are appeals to the regular ordering of our sensory ideas.

As a result, Berkeley's instrumentalism is one about signs. Instead of supposing that the objects of mathematical and scientific inquiry are metaphysically real entities, we need to acknowledge that the objects of study are *signs* for ideas and their relations. Berkeley gives us a clear example in the 19th question of the *Analyst.*

> *Qu.* 19 When it is said or implied, that such a certain line delineated on paper contains more than any assignable number of parts, whether any more in truth ought to be understood, than that it is a sign indifferently representing all finite lines, be they ever so great. In which relative capacity it contains, *i.e.* stand for more than any assignable number of parts?[16]

The calculus, strictly speaking, concerns a particularly useful way of manipulating signs, just as the square roots of negative numbers help us more accurately predict future experiences. No matter how useful the sign it might be, for Berkeley no one ought to seriously think that there is such a 'thing' as *'i.'*

Theories, however, are complicated by the fact that we do not tend to separate active from passive elements in our thinking about the world. For Berkeley, we cannot have ideas of active things, since ideas are passive and represent only by likeness. Only an idea can be like an idea, hence we cannot represent (with ideas) objects that are active, or have causal powers, and so on. Berkeley has a clean way of dealing with such complications, however, and he borrows it from his mature metaphysics. We know spirits only by their effects. "Such is the nature of *spirit* or that which acts, that it cannot be of itself perceived, but only by the effects which it produceth."[17] The same point applies consistently for everything that has an essentially active nature, including our own volitions. "We see no variety or difference betwixt the Volitions, only between their effects."[18] We ought, then, to expect the same for objects of study in the sciences – and that is exactly what we find. Alleged

---

[13] Compare AN 22, 78. For one example where Berkeley mentions true claims that might be useless or vain, see ALC 308.

[14] ALC 293–295.

[15] See PHK 131, where Berkeley argues that denying the literal existence of infinitesimals does not negate the usefulness of geometry and mathematics.

[16] AN Qu. 19, *Works* IV, 97.

[17] PHK 27.

[18] PC 788.

active objects, like forces, are known only through their effects. One might think that motion is an exception, since motion appears active yet Berkeley says that we can have ideas of motion. He denies, however, that motion is really active or an activity, despite how we talk about it. "Motion, though in metaphysical rigour and truth a passion or mere effect, yet in physics passeth for an action."[19] Although motion might be said to be an action, in fact it is a "mere effect." Emphasizing the perceivable effects of scientific posits also helps Berkeley deal with the potential concern that there is a principled difference between concepts that are not in principle perceivable (like forces) and those that are (the movement of the earth). In both cases, it is the *effects* that matter (and those are always perceivable because they are in fact perceived), and thus whether the *posit* is itself in principle perceivable is not really a worry on Berkeley's analysis. We have, then, a doctrine that is consistent between Berkeley's early and late works. We do not have ideas of volition, force, or activity, but we do use *signs* to represent them through their effects, which are sets of ideas, presumably well-ordered.

We accordingly should think of his views in terms of modeling sensory experiences. A theory for Berkeley is to be judged foremost (but not only) by its empirical adequacy – its ability to allow us to predict what sensory ideas we will have given current conditions and an understanding of the regularities we find in the perceived world. "If I mistake not, all sciences, so far as they are universal and demonstrable by human reason, will be found conversant about signs as their immediate object, though these in the application are referred to things."[20] Science is about *signs* and their connections. If we abandon thinking about mathematics as about *actual* infinities and infinitesimals (i.e. about things that allegedly underlie the experiences we have), then "those and the like objections vanish, if we do not maintain the being of absolute external originals, but place the reality of things in ideas ...."[21] Science and mathematics amount in reality to nothing more than organizing and making sense of the experiences we have.[22]

Berkeley is an instrumentalist; he believes that these signs are valuable because of their usefulness – their predictive accuracy – rather than being faithful depictions of some underlying (beyond the train of ideas) reality. Thus far, all is well. Yet one might ask at this point what limits the *acceptability* of a theory. Is it mere utility? Not for Berkeley. In order to have an acceptable theory, it must be the case the objects being modeled are also antecedently logically possible. This constraint, as it turns out, is the same limitation Berkeley places on ideas in general. Just as every idea must have (or rather be) a possible object, every theory must model something that is logically possible (though not necessarily actual). One cannot have a *useful* sign for an impossible object. The constraint, I suggest, is perfectly reasonable. We now believe that science must involve only the observable world. There is no

[19] S 161.

[20] ALC 305.

[21] 3D 258.

[22] See PHK 58.

'science' for the soul, for instance, since by hypothesis it is not observable. Similarly, there is no science for the impossible, which is likewise unobservable. This second constraint, in conjunction with other textual moves, allows us to read Berkeley as a consistent basic instrumentalist about mathematics and science, and allows us to explain certain puzzles in the texts. In short, science is a useful method for organizing possible experience. For Berkeley, if a posit or theory is not useful, *or* it is not possible, then it is not part of science.

2.3 The Constraint of Possibility

Precedents exist for the kind of view I attribute to Berkeley. Leibniz arguably shifts to a phenomenal understanding of matter in part because of problems like the angle of the tangent, where the mathematical formalisms describe something that he thinks is 'literally' impossible in an actual material world. Berkeley, I believe, has a similar view. One of his principal objections against the existence of abstract ideas is that such ideas are metaphysically impossible, since they posit (or require) indeterminate particular beings.[23] Berkeley extends this kind of analysis to the sciences.

If we start with the simplest cases of instrumentalism in his writings, we can see that Berkeley wants to adopt theories on the basis of their usefulness rather than their accuracy in describing a mind-independent reality. The point of walking through one or two well-accepted cases is to confirm that they are examples in which the posited objects are also logically possible, even if obscure. In *De Motu* he is most clear:

> Force, gravity, attraction, and terms of this sort are useful for reasonings and reckonings about motion and bodies in motion, but not for understanding the simple nature of motion itself or for indicating so many distinct qualities. As for attraction, it was certainly introduced by Newton, not as a true, physical quality, but only as a mathematical hypothesis. Indeed Leibniz when distinguishing elementary effort or solicitation from impetus, admits that those entities are not really found in nature, but have to be formed by abstraction.[24]

Force, as Berkeley tells us in *De Motu* Section 5, is an occult quality. As a sign, however, it is describing something not only possible, but actual. That is, 'force' is our sign signifying sets of certain connected experiences. The error is to think that there is something – force – instead of recognizing that the word is only an instrument of the mind. Berkeley employs this language of the mind using 'instruments' significantly in the *Siris.*

Unperceived motion is another particularly clear case. At *Principles* 58 Berkeley engages the objection that his immaterialist version of science runs counter to established claims like *the earth moves.* The charge is that because we do not

[23] See my *Idea and Ontology* (University Park, Penn State University Press, 2008), chapter 8. See IPHK 23 for the emphasis on the impossibility of *determinate* abstract ideas.

[24] DM 17.

actually perceive the earth moving, Berkeley's theory must be incompatible with the scientific claim that it does.

> Tenthly, it will be objected, that the notions we advance, are inconsistent with several sound truths in philosophy and mathematics. For example, the motion of the earth is now universally admitted by astronomers, as a truth grounded on the clearest and most convincing reasons; but on the foregoing principles, there can be no such thing. For motion being only an idea, it follows that if it be not perceived, it exists not; but the motion of the earth is not perceived by sense. I answer, that tenet, if rightly understood, will be found to agree with the principles we have premised: for the question, whether the earth moves or no, amounts in reality to no more than this, to wit, whether we have reason to conclude from what hath been observed by astronomers, that if we were placed in such and such circumstances, and such or such a position and distance, both from the earth and sun, we should perceive the former to move among the choir of the planets, and appearing in all respects like one of them: and this, by the established rules of Nature, which we have no reason to mistrust, is reasonably collected from the phenomena.[25]

To say that the earth moves is shorthand for observing that certain ideas we perceive are well-ordered in regular ways. Here again, there is nothing impossible about motion, and thus Berkeley's instrumentalism is fairly straightforward. We may safely conclude that the clearest and most uncontroversial cases of his instrumentalism satisfy the constraint of possibility.

Evidence of the added possibility constraint in more challenging cases appears in the *Analyst*, where Berkeley engages Newton's theory of fluxions in his calculus. Berkeley challenges Newton's theory not on the grounds that fluxions are not useful, but rather by arguing that the theory asks us to frame ideas that are inconceivable in virtue of their being impossible.

> The further the mind analyseth and pursueth these fugitive ideas the more it is lost and bewildered; the objects, at first fleeting and minute, soon vanishing out of sight. Certainly in any sense, a second or third fluxion seems an obscure mystery. The incipient celerity of an incipient celerity, the nascent augment of a nascent augment, i.e. of a thing which hath no magnitude: take it in what light you please, the clear conception of it will, if I mistake not, be found impossible; whether it be so or no I appeal to the trial of every thinking reader. And if a second fluxion be inconceivable, what are we to think of third, fourth, fifth fluxions, and so on without end?[26]

Berkeley goes on to oppose the Leibnizian concept of infinitesimals on the same grounds. Eventually he rejects the new calculus for a rather specific reason.

> But, notwithstanding all these assertions and pretensions, it may be justly questioned whether, as other men in other inquiries are often deceived by words or terms, so they likewise are not wonderfully deceived and deluded by their own peculiar signs, symbols, or species. Nothing is easier than to devise expressions or notations, for fluxions and infinitesimals of the first, second, third, fourth, and subsequent orders, proceeding in the same regular form without end or limit &c. or dx. ddx. dddx. ddddx. &c. These expressions indeed are clear and distinct, and the mind finds no difficulty in conceiving them to be continued beyond any assignable bounds. But if we remove the veil and look underneath, if, laying

[25] PHK 58.

[26] AN 4.

aside the expressions, we set ourselves attentively to consider the things themselves which are supposed to be expressed or marked thereby, we shall discover much emptiness, darkness, and confusion; nay, *if I mistake not, direct impossibilities and contradictions.*[27]

When we use words as signs for objects that are themselves impossible, we have grounds for rejecting any theory that employs them, independently of whether there might be some usefulness to be had. The point is that instrumentalism does not *apply* except in cases where we already have a viable science. Note that Berkeley is attacking the *words* (expressions) used by Newton and Leibniz only insofar as they are *signs* for objects to be perceived by minds. I conclude that there are excellent grounds for thinking that Berkeley is an instrumentalist, provided we recognize that his instrumentalism is tempered by an additional metaphysical constraint that he imposes on the acceptability of theories that extends beyond their mere usefulness.

Having laid bare the basics of my reading of Berkeley's instrumentalism, there are two immediate challenges to its tenability. First, large portions of the *Siris* appear to run directly counter to my general claim that Berkeley is a consistent basic instrumentalist. Second, Jesseph has argued that Berkeley only endorses a weaker form of instrumentalism with respect to mathematics (from the *Analyst* in particular). I believe both concerns can be met, and I will engage each in turn.

2.4 The Complication of *Siris*

In her excellent article on Berkeley's instrumentalism, Lisa Downing notes what she takes to be his commitment to the existence of corporeal particles in the *Siris* and asks why Berkeley is an instrumentalist about force but a realist about aetherial particles, at least in that late work.

Moreover, Berkeley seems to assume the existence of many of the theoretical entities he describes, including particles of aether .... In *De Motu* and *Alciphron*, Berkeley in effect develops a form of instrumentalism which he applies to Newton's mechanics. Why, one might well wonder, do corpuscles not receive a similar treatment in *Siris*, why does Berkeley not extend his instrumentalism to aetherial particles, for example?[28]

My answer, in short, is that he *does* extend his instrumentalism to *Siris*. Justifying my response, however, first requires some textual analysis.

I start with some brief observations about the *nature* of *Siris*; it is, by all accounts, an unusual text unlike the rest of Berkeley's published writings. Exactly what scholars of early modern philosophy ought to do with it has generated a flurry of activity early across the twentieth-century, with more recent commentators like Gabriel Moked arguing that Berkeley essentially gave in and changed his mind,

[27] AN 8, my emphasis.

[28] Downing, "*Siris* and the Scope of Berkeley's Instrumentalism," 281.

embracing his own form of the new corpuscularian philosophy.[29] It lies outside the scope of this paper to engage Moked's book, but I shall present a view here that rejects his basic thesis. I take *Siris* to be a work where Berkeley is trying to insinuate and make more palatable some of his unique metaphysics into the mainstream views of the day. The book starts with a defense of the curative powers of tar-water – hardly a sage introduction to a weighty and serious philosophical tome. In the work, however, one can see Berkeley trying to make room for his own metaphysical claims within the corpuscularian and chemical traditions of his day. I shall endeavor to make this clear as I engage Downing's analysis.

Downing argues that Berkeley's attitude towards particles (especially aetherial particles, or pure fire) is "straightforwardly realistic," implying that Berkeley admits the existence of corporeal particles that are neither ideas nor collections of ideas.[30] As a result, *Siris* presents a special problem for people trying to understand the nature of Berkeley's instrumentalism. Downing concludes that Berkeley is an instrumentalist about dynamics and a realist about corpuscles in the *Siris*, but has a principled reason for excepting corpuscles from his general views. This reason saves the consistency of his instrumentalism throughout his works. It should be noted, however, that Downing's reading leaves us with the unpalatable result that Berkeley was, at least to some degree, not absolutely consistent in his metaphysics even if he was with respect to his instrumentalism.

I freely admit that many passages in *Siris* initially read as if Berkeley were an unreflective realist about particles, but even Downing notes how hard it is to pull out a consistent interpretation of Berkeley as a realist throughout the text. We have passages that superficially seem to indicate that aether or pure fire is corpuscular in nature, as in the following:

> We are not therefore seriously to suppose, with certain mechanic philosophers, that the minute particles of bodies have real forces or powers, by which they act on each other, to produce the various phenomena in nature. The minute corpuscles are impelled and directed, that is to say, moved to and from each other, according to various rules or laws of motion.[31]

The contrast here with dynamical forces appears striking. Right after denying the reality of forces Berkeley goes on to speak of corpuscles as if they really existed. He even speaks of light as corporeal.

> But it is now well known that light moves; that its motion is not instantaneous; that it is capable of condensation, rarefaction, and collision; that it can be mixed with other bodies, enter their composition, and increase their weight (Sects. 169, 192, 193). All which seems sufficiently to overthrow those arguments of Ficinus, and shew light to be corporeal.[32]

[29] Gabriel Moked, *Particles and Ideas: Bishop Berkeley's Corpuscularian Philosophy*, Oxford: Clarendon, 1988, 25. For a further discussion of the varying views, see Wilson, "Berkeley and the Microworld," 37–39.

[30] Downing, "*Siris* and the Scope of Berkeley's Instrumentalism," 283.

[31] S 235.

[32] S 207.

Berkeley does often write as if aether is composed of corporeal bodies that interact with other corporeal beings.[33] I have no intention of denying how the texts readily appear.

The problem, however, is that Berkeley also says things in *Siris* that are flatly at odds with the corpuscularian understanding of aether. Downing is smartly aware of these passages and lists a few of them. Here are some key passages, the first three of which Downing cites as well.

> The pure fire is to be discerned by its effects alone…. (S 190)

> …fire is a subtle invisible thing, whose operation is not to be discerned but by means of some grosser body, which serves… for a vehicle to arrest and bring it into view… (S 197)

> The pure invisible fire or aether doth permeate all bodies, even the hardest and most solid, as the diamond. (S 200)

> This pure spirit or invisible fire is ever ready to exert and shew itself in its effects…. (S 157)

If the aether is *invisible* and known *only* by its effects, then it is not perceivable in the way that bodies are within Berkeley's system. In short, the very corpuscles about which Downing believes Berkeley is a realist at a minimum do not seem to be real in the same way that bodies are according to materialist theories. What are we to do with these passages? Downing concludes that "there does seem to be unresolvable tension between Berkeley's championing of the aether in *Siris*, and the particular *esse est* [sic] *percipi* position defended in his early works. It appears that Berkeley, whether knowingly or not, has relaxed his earlier criterion for actual existence."[34] In sum, Downing's account of Berkeley's instrumentalism leaves us with an unpleasant inconsistency in his metaphysics.

I want to suggest that something else is going on, based on a reading of *Siris* as it unfolds towards those sections where he discusses aether. In the sections starting around 140, Berkeley argues that air is key to life, but only because it acquires a property that makes it life giving. The attraction of an *active* subtle substance called fire, aether, light, or vital spirit is required (S 147). At this point Berkeley starts discussing the nature of aether, and when he does so he is careful to emphasize that the strict nature of aether is known only by its effects. "This pure spirit or invisible fire is ever ready to exert and shew itself in its effects (Sect. 152), cherishing, heating, fermenting, dissolving, shining, and operating in various manners, where a subject offers to employ or determine its force."[35] Note that this spirit is active. Just like minds from the early metaphysical works, we only know active things by the

[33] Compare S 162. Downing claims that S 207 and 162 show that Berkeley's aether have some 'determinate size, shape, weight, etc.,' although as I shall argue I think she has over-read the passages.

[34] Downing, "*Siris* and the Scope of Berkeley's Instrumentalism," 289–290.

[35] S 157.

effects they produce.[36] Berkeley's descriptions of pure fire or aether are consistent. Pure fire is an "active principle" (reminiscent of 3D 233–4 where Philonous describes himself *qua* mind as 'a thinking active principle'). Berkeley even apparently connects his account of aether to the nature of spirits/minds in *Siris* Section 159.

> No eye could ever hitherto discern, and no sense perceive, the animal spirit in a human body, otherwise than from its effects. The same may be said of pure fire, or the spirit of the universe, which is perceived only by means of some other bodies, on which it operates, or with which it is joined. What the chemists say of pure acids being never found alone might as well be said of pure fire.[37]

By the time we have reached Section 160, we have the core account in place. Berkeley writes, "The mind of man acts by an instrument necessarily."[38] Before Berkeley introduces the discussions of aether he is telling us that by 'aether' he means a *sign*, itself invisible, of experiences we have in the phenomenal world. He does not want to emphasize the fact because he wants the work to have the sort of appeal to the vulgar that his earlier, more avowedly philosophical, works did not. He posits aether as an instrument the mind uses to organize certain kinds of regular experiences.

As a result, there is no need to provide an account that squares Berkeley's instrumentalism about dynamics with his realism about corpuscles because he is an instrumentalist about both. Recall that Berkeley's instrumentalism does not deny that ideas (and their orderings) are real; he denies that there is an *additional* underlying reality to which one must appeal in order to do science. One may *talk* about corpuscles and use all sorts of signs, but Berkeley is quite clear that, in reality, there is nothing beyond the effects we perceive. Thus his mentions of corpuscles (and similar items) are meant to be understood as signs used to explain regularities in the ordering of our ideas. Berkeley is 'speaking with the vulgar' in order to connect his other views with the popular intellectual currents of his day, including not only the new corpuscularian science, but also chemistry and views like the Great Chain of Being to which he alludes late in *Siris*.[39] Berkeley's invocation of the chain of being strikes me as evidence that he is only trying to make his philosophical system more palatable by showing it to be consistent with already well accepted views.

To make instrumentalism work with respect to aether, we need only note first, that the words like 'aether' he uses are signs for sets of effects (ideas or experiences) that are useful, and second, that these posits are logically possible (hence science can apply to them). From the previous cited passages I think it quite clear that Berkeley asks us to understand pure fire and aether only through their observable effects. I can find no better explanation for why he carefully inserts

---

[36] I am leaving aside considerations of intuitive or notional knowledge, which Berkeley does not invoke in these specific discussions in any event.

[37] S 159.

[38] S 160.

[39] See S 303: "There runs a chain throughout the whole system of beings .... The meanest things are connected with the highest."

and repeats claims about the active nature of aether in a work clearly designed to have appeal to an educated but lay audience.

I also want briefly to comment on Downing's own theory about the scope of Berkeley's instrumentalism based on his distinction between ways in which theories are generated. I think her analysis here is essentially correct, and so my aim is only to demonstrate how her distinction is compatible with my reading of Berkeley's instrumentalism.

Downing argues that Berkeley distinguishes between two types of scientific method, an inductive method based on sensory experiences and a hypothetic-deductive method.[40] The key passage upon which she relies appears in *Siris*.

> It is one thing to arrive at general laws of nature from a contemplation of the phenomena, and another to frame an hypothesis, and from thence deduce the phenomena. Those who supposed epicycles, and by them explained the motions and appearances of the planets, may not therefore be thought to have discovered principles true in fact and nature. And, albeit we may from the premises infer a conclusion, it will not follow that we can argue reciprocally, and from the conclusion infer the premises.[41]

We are 'naturally' realists about the results of induction, but hypotheses should be handled with care and only evaluated on the basis of their usefulness. In particular, we should not infer the existence of any posited entities, accepting as real only those things we actually perceive. Newtonian dynamics is hypothetical (and Downing cites a passage from Newton's *Principia* that confirms her assertion), but the science of aetherial corpuscles Berkeley advances in the *Siris* is experiential and inductive. As a result, Downing has a principled basis on which to argue that Berkeley is an instrumentalist about some things (those involving the hypothetical method), but not others (inductive methods).

About the distinction and that Berkeley held it I am convinced. He separates two methodologies in the conduct of science. The issue now is whether Berkeley's claims about the alleged reality of aetherial particles is evidence that he is not an instrumentalist. If my supposition that words like 'aether' and 'pure fire' are simply signs for collections of well-ordered experiences is correct, then we might question whether the use of such a method automatically excludes instrumentalism. Nothing in the above passage (or elsewhere in *Siris*) requires that we understand Berkeley to think that we must believe in anything beyond the ideas we perceive. Inducing and anticipating ideas from previous experiences is a separate method, but it is not one that *requires* that we posit 'real' entities that underlie those ideas. We use signs as useful 'instruments' of the mind (remembering *Siris* 160), even though those signs do not necessarily pick out mind-independent things. There is no reason to think Berkeley must be appealing to an underlying causal order here. In fact, his emphasis seems clearly otherwise. At least part of the *point* of the discussion of aether in *Siris* is to explain *why* tar water is such a potent curative.

---

[40] Downing, "*Siris* and the Scope of Berkeley's Instrumentalism," 293–294.

[41] S 228.

That aim is instrumental. Berkeley's alleged realism about particles thus has to be understood under two constraints. First, Berkeley *tells us* what he means by those words (e.g. 'aether') in the *Siris*, namely that they are known only by effects. Second, we must bear in mind that the *Siris* is an unusual work arguably designed to make his core metaphysical views more acceptable to the educated lay audience of his day. These would be readers with at least a passing familiarity with curatives and alchemy, as well as other intellectual currents – many of which Berkeley discusses. In short, *Siris* is not a sober 'in truth and strictness' hard work of 'pure' philosophy. We thus need to allow Berkeley some latitude in his expression given his larger aims.[42]

2.5 The Challenge of Geometry

A second challenge awaits my account. Douglas Jesseph has argued that Berkeley retreats from a full-blown instrumentalism with respect to his discussion of the calculus in the *Analyst*. Berkeley there argues that the conclusions of the Newtonian and Leibnizian calculi are correct, but the methodologies are in serious error. Jesseph rightly asks why an instrumentalist would care whether the methodologies were right, so long as the theories were useful. I am in the odd position here of agreeing with virtually all of Jesseph's claims. He even notes the roughly same constraint (of possibility) that I articulated earlier. That said, I think Jesseph draws a less charitable conclusion about Berkeley's instrumentalism even as his analysis of Berkeley's philosophy of mathematics and science is otherwise excellent. If we ignore the constraint of possibility, then I believe Jesseph is probably right: Berkeley is no instrumentalist about mathematics *simpliciter*. When we add the constraint, however, I think we can fairly attribute to him a consistent form of instrumentalism.

Consider Berkeley's attack on the calculus. He explicitly admits that the conclusions are true, choosing instead to focus on the underlying methodologies.

> I have no controversy about your conclusions, but only about your logic and method. How you demonstrate? What objects you are conversant with, and whether you conceive them clearly? What principles you proceed upon; how sound they may be; and how you apply them? It must be remembered that I am not concerned about the truth of your theorems, but only about the way of coming at them; whether it be legitimate or illegitimate, clear or obscure, scientific or tentative.[43]

In short, Berkeley claims that the methodologies are flawed, but happen through the good fortune of 'compensating errors' to yield true conclusions. I am not interested

[42] I am absolutely *not* arguing that Berkeley is being disingenuous in *Siris*. Instead, my claim is that the work is self-reflectively less explicit about issues of presentation in order to make the claims advanced in it more accessible to readers who might otherwise find the unadulterated metaphysics of immaterialism less than palatable.

[43] AN 20.

here in the quality of his unusual diagnosis, which Jesseph has ably shown to otherwise be suspect. What matters for my present purpose is whether this sort of analysis precludes Berkeley from being a consistent instrumentalist.

Jesseph argues that "Berkeley's struggle to use compensating errors in accounting for the success of the calculus is also part of his rejection of a thorough-going mathematical instrumentalism. His effort to explain away the paradox of true conclusions drawn from false premises is exactly the kind of work an instrumentalist need not bother with."[44] Exactly right, provided one is already convinced that we are actually *doing* scientific work. If, however, we add the constraint of possibility, we might see Berkeley's concern as the following. If the methodologies Newton and Leibniz employ invoke impossible entities, then there is a sense in which *they are not doing science.* Jesseph diagnoses the difference between works like *De Motu* and *Alciphron* on the one hand, and the *Analyst* on the other, as the difference between metaphysical and logical critiques. But if Berkeley's aim in the *Analyst* is to preserve the conclusions of the calculus by making it a science when it otherwise would not be, then perhaps we have a more charitable position to attribute to him.

At the conclusion of his discussion of the compensating errors thesis, Berkeley writes:

> This hint may perhaps be further extended, and applied to good purpose, by those who have leisure and curiosity for such matters. The use I make of it is to shew, that the analysis cannot obtain in augments or differences, but it must also obtain in finite quantities, be they ever so great, as was before observed.[45]

As Jesseph aptly notes, the point is to replace the use of infinitesimals with something that is possible: his "finite quantities, be they ever so great."[46] We saw earlier (AN 8, quoted above) what the problem was: Berkeley thinks that infinitesimals and fluxions are not possible entities. As a result, I speculate that he believes that the calculus, as presented, is not really a science or proper mathematics at all. To the degree to which it uses abstract (i.e. impossible) ideas, it *cannot* be.

Jesseph himself later admits something like what I have called the constraint of possibility I am here invoking.

> Thus, only those terms that have observable content 'taken in concrete' are, in Berkeley's view, properly scientific: 'In illuminating nature it is vain to adduce things which are neither evident to the senses nor intelligible to reason. Let us therefore see what sense, what experience, and lastly what reason resting up them recommend' (*De Motu* §21). Clearly, Berkeley's instrumentalism does not extend so far as to permit theories whose terms lack all experiential content.

In essence, my disagreement with Jesseph is rather minor. He notes the constraint and uses it to conclude that Berkeley is not an instrumentalist whereas I want to fold the constraint into Berkeley's thinking about what it is to do science.

---

[44] Jesseph, *Berkeley's Philosophy of Mathematics*, 213.

[45] AN 29.

[46] Jesseph, *Berkeley's Philosophy of Mathematics*, 212–213.

2.6 Metaphysical Instrumentalism

Central to my argument in this paper is my here unsupported contention that one ought not distance Berkeley's metaphysics from his thinking anywhere else. His unusual commitments to stark metaphysical principles permeate his philosophy. Thus, in order to be charitable to Berkeley and his views we need to interpret his claims in light of these principles. He argues at length against the existence of abstract ideas on the grounds that they are impossible entities. Applying this bit of his metaphysics to this thinking about science yields a more charitable reading of him. By taking seriously how Berkeley unpacks and introduces the concepts of aether, light, etc. in *Siris* and by adding the metaphysical constraint of possibility to what counts as science (including mathematics), we find ourselves with a way to resolve some of the tensions that appear in the Berkeleian corpus with respect to both his instrumentalism and his metaphysical system.

Berkeley holds that science properly speaking is concerned with possibilia and their use. If one attempts to apply scientific methodology to impossible (contradictory) entities, then one is not even doing science. And when applying the methods of science to the world one does not require recourse to a 'deeper' causal reality underneath the ordering of ideas set down by God. That, I suggest, may make him a consistent instrumentalist.

Chapter 3
Causation, Fictionalism and Non-Cognitivism: Berkeley and Hume

P.J.E. Kail

'Non-cognitivism', a creature of the twentieth century, is nevertheless associated with David Hume and George Berkeley, two figures of the eighteenth, often in relation to their ethical theories. I am not here concerned with the history of this association, nor with the application (or misapplication) of 'non-cognitivism' to Hume's or Berkeley's views on ethical thought and talk. Instead I want to compare their treatments of the causal relation, and probe the extent to which each can be understood as offering, albeit in embryonic form, a non-cognitivist account of it. The purpose of this exercise, however, is not taxonomic. It is questionable whether such an exercise would be of much weight or interest, even if we were to ignore the obvious problem of anachronism. For the identity of 'non-cognitivism' is itself contested in contemporary philosophy, and so there is, in fact, no settled way to determine what exactly is 'non-cognitivism'. Because any definition of 'non-cognivism' would, of necessity, foreclose on a host of difficult matters I shall not offer any such definition. I shall instead examine the accounts of Hume and Berkeley with an eye to what kinds of considerations might invite the label 'non-cognitivism' and in what senses that label might be appropriate.

3.1 I

Let us begin with Berkeley. Non-cognitivism and its variants are associated with his philosophy predominantly because of some of the remarks he makes about the nature and function of language in the Introduction to the *Principles of Human Knowledge* (PHK), in the draft of that introduction, and in *Alciphron*.[1] These remarks

[1] For discussion see e.g. Belfrage (1986) and (1987); Berman (1986) and (1994); Jakapi (2002) and Jakapi (2003); Williford (2003).

P.J.E. Kail (✉)
St. Peter's College, Oxford University, New Inn Hall St., OX1 2DL, Oxford, England
e-mail: sfop0159@herald.ox.ac.uk

S. Parigi (ed.), *George Berkeley: Religion and Science in the Age of Enlightenment*, International Archives of the History of Ideas/Archives internationales d'histoire des idées 201, DOI 10.1007/978-90-481-9243-4_3,

and their relation to non-cognitivism are connected in the literature with the topics of the Christian mysteries and ethical discourse but not, to my knowledge, with his views on natural causation. But as we shall see, the connection between these remarks and his views on natural causation is very close indeed.[2]

What are these remarks? In PHK introduction §19 Berkeley argues against the 'received opinion' that 'language has no other end but the communicating our ideas, and that every significant name stands for an idea'. Notice that two ideas fall under Berkeley's suspicion here. The second mentioned that every name suggests an idea, and Berkeley's immediate response to this received opinion is to note that the names used in algebra need not suggest an idea in each and every use. The first idea that Berkeley questions, and which is our concern, is that the sole use of language is the communication of ideas. In PHK introduction §20 Berkeley adumbrates other ends of language, including 'the raising of some passion, the exciting to, or deterring from an action, [and] the putting the mind in some particular disposition'. In such cases, he adds that the communication of ideas is 'in many cases barely subservient, and sometimes entirely omitted'. It is these remarks that invite the term 'non-cognitivism'. But do they really support such an attribution?

If we understand by that term what has become known as 'non-factualism', then I think there is no evidence in these remarks for non-cognitivism so construed. By 'non-factualism' I shall understand the idea that, despite having an apparently fact-stating linguistic shape, the semantic content of the discourse cannot be construed as being in the business representing states of affairs and is therefore[3] incapable of being true or false. One *might* think Berkeley's remarks imply something like non-factualism on the following basis: any statement capable of being true or false must express a representational content, and any such content requires ideas. However, there are ends of language that have a function other than conveying an idea, and so it seems to follow that representation is not involved and the semantics here must be non-factual. This, however, won't do. It is perfectly possible to perform different speech acts with sentences, such as directives, comissives and the like, which are not primarily a matter of communicating ideas (i.e. those that are not 'constatives', as J. L. Austin put it) and yet this puts no pressure on the idea that there is a propositional content at their core.[4] I can say to my wife 'Rome has many restaurants' with the intention of commending it to her as a place to visit, and yet it is a straightforward matter of fact that Rome has many restaurants. It seems to me that the Introduction remarks, taken by themselves, imply nothing more than this relatively homely thought. So where might a non-cognitivist reader go from here?

If we are to have a thesis more interesting that the fact that we can do more with words that state things (not, of course, that this is itself uninteresting), a common starting

[2]For my first attempt and understanding this connection, see my (2007b).

[3]The emergence of minimalism makes this a questionable 'therefore'. Recall, as I said in the introduction, that the identity of 'non-cognitivism' is contested, and so it is dangerous to begin with too rigid a conception of it.

[4]Austin (1962).

point should point to something about a particular *area* of discourse that invites non-cognitivism. One might then argue on independent grounds that for some particular area of discourse it is impossible to provide genuine propositional content. This would establish non-factualism, and one could then connect the remarks in the Introduction to the relevant area of discourse, and establish a non-cognitivist semantics. Roughly, because we cannot be using the language to represent, we must therefore be doing something else when we use it. Alternatively, one might argue that since the primary purposes of the predicative uses of the distinctive terms in the discourse are not fact stating, one should view the semantics as non-cognitive.[5] A third option, which I shall pursue here, is to see non-cognitivism as the correct *psychology* for the area, such that acceptance in the area of discourse consists in the possession of some attitude (such as a disposition to act) other than belief. This would keep the semantics factual, but insist that what we *do* with such contents does not consist in believing the world to be how they represent it to be. This general philosophical position, known as 'fictionalism', is a newcomer on the philosophical scene, and the subject of much recent critical discussion.[6] Here I consider it in connection with Berkeley. Applying the label of 'fictionalist' to the Bishop's philosophy of course carries with it the threat of anachronism, but, one should remark, no more so than applylling the label of 'non-cognitvist' does. Be that as it may, it serves as a useful exegetical heuristic.

3.2 II

The remarks we have been considering occur in the Introduction to the *Principles* and so one might reasonably expect them to play some role in the doctrines espoused in the body of that work. Without prejudice to other areas of Berkeley's thought that might be informed by such remarks (such as his treatment of mathematics), it is with respect to natural causation that they seem most pertinent. Berkeley believes that ideas are themselves passive, and that the only genuine power is volitional. This implies that the relation among ideas is not a causal one, if that is understood in terms of one sensible object (an idea) having the power to bring about another. This, however, appears to impute to ordinary thought systematic error in the endemic use of causal verbs like 'push', 'burn', 'scratch' and a host of others. This is clearly an unwelcome consequence for a philosopher who wants to align his philosophy as much as possible with commonsense.

Berkeley's response, as we shall see, is to recommend that we keep the talk, despite the falsity of beliefs. But this is not the only accusation of the imputation of massive error that Berkeley considers, and it is instructive to consider how his responses to these two objections differ. At PHK §54 he canvasses the idea that

---

[5] I argued this in connection with Berkeley and natural causation in my (2007a), but I am no longer committed to the semantic thesis.

[6] See, for example, the contributions to Kalderon (2005).

immaterialism implies that we are systematically mistaken in our belief in physical objects. Berkeley's response is to deny that any genuine thought, and hence belief, can be expressed by physical object terms, since the notion of a sensible object without the mind – a physical object – implies a contradiction. Such a response is congenial to the non-factualist, since Berkeley is in effect arguing that any sentence using physical object predicates is incapable of expressing a genuine thought and so incapable of being true or false. But this is not the response offered in the case of causal predicates. Instead, he famously says that we 'ought to think with the learned, and speak with the vulgar' (PHK §51). With respect to thinking with the learned, Berkeley implies in the next section that uses of the causal predicates are truth-apt but false. For he says that they may be retained despite 'how false soever they may be, if taken in a strict and philosophical sense' (PHK §52). So, unlike the case of material object predicates, it seems that some genuine thought can be expressed by 'x causes y' when applied to sensible objects, but any such thought is false. Non-factualism does not, therefore, appear to be Berkeley's position. Why then ought we 'speak with the vulgar'?

In connection with this, Berkeley states that in the 'ordinary affairs of life, any phrases may be retained, so long as they excite in us proper sentiments, or dispositions to act in such a manner as is necessary for our well-being' (PHK §52). This manifestly echoes the remarks in the Introduction where the ends of language include 'the raising of some passion, the exciting to, or deterring from an action, [and] the putting the mind in some particular disposition'. Effectively, then, we are allowed to speak with the vulgar because despite the falsity of causal predications they serve these other, non-representational, ends of language.

Talk of exciting proper sentiments and dispositions, however, is highly schematic. Can any more detail be provided? Let us begin by noting a distinction drawn in contemporary fictionalism between 'hermeneutic' and 'revolutionary' fictionalisms.[7] The first view has it that in fact putative representational claims, and the acceptance of them, are generally geared primarily to ends other than representation. Revolutionary fictionalism, on the other hand, views ordinary discourse as typically engaged in representational practices, but suggests that we *ought* to treat such sentences as serving ends other than representation. This is because as the discourse stands it is, in one way or another, deeply problematic. Now for any area of discourse we shall need some strong reasons for *either* interpreting our discourse as the hermeneutic fictionalist proposes *or*, if one is revolutionarily inclined, provide both a reason why the discourse is problematic *and* an account of what purposes the fiction ought to serve.

In the light of this, it seems, that Berkeley's proposal is revolutionary – a claim about how we *should* treat ordinary language discourse – and the motivation behind the revolution is the fact that taking such language literally implies widespread error. The remaining issue, namely, the matter of what purpose the fiction is supposed to

[7] See Stanley (2001).

serve, is a great deal more complex, but nevertheless integral to Berkeley's system. And it provides an answer to the question of what 'proper sentiments, or dispositions to act' causal talk is supposed to elicit. To see this, recall Berkeley's celebrated view that the relations among sensible objects are primarily *semantic*. The natural world is constituted by the language of God, and the business of science is its interpretation. Relatively fundamental to this picture is the semantic relation of *sign* and *signified*. We are used, in this connection, to making the familiar distinction between *natural* and *conventional* relations of sign and signified. A *natural* sign gets its status as such inasmuch as there is a causal relation between the sign and that for which it is a sign. Thus spots are signs of measles in virtue of the fact that spots are a causal consequence of that disease, wherein the effect is a sign of a cause. Conversely, clouds are signs of rain because clouds bring about precipitation. Conventional meaning, however, is a matter of an arbitrary connection between some sign (such as 'cat) and some object (cats). But this distinction is not sustainable on Berkeley's system. Because Berkeley denies that there are causal relations among ideas, all the sign/signifier relations are conventional. So what we take to be causal relations are in fact conventional relations of sign and signified.

It is here that Berkeley's revolutionary proposal can get some purchase. For the language of God, which constitutes the world, is not in place simply to communicate facts. The world is ordered providentially; that is its arrangement is conducive to the well being of humanity. So in order to interpret properly this language one needs to know not merely what is as sign for what, but also grasp what *action* is appropriate in the circumstances. This point itself relates to the remarks in the PHK introduction. For recall that the ends of language include the 'raising of some passion, the exciting to, or deterring from an action, [and] the putting the mind in some particular disposition'. Since natural events constitute a language, and God's utterances relate to the end of human well being, then some of those utterances are intended to elicit in the audience the appropriate behavioural stance. So, for example, the relation between heat and pain can be construed as the heat's being a warning of possible injury, and hence an utterance in God's language that has the aim of deterring action. But now, if I tell you that heat will *cause* your skin to burn, or that smoking *causes* cancer, such statements might themselves put you into the appropriate behavioural stances. This is the key to Berkeley's revolutionary fictionalism. For he writes that causal expressions 'may be retained, so long as they excite in us proper sentiments, or dispositions to act in such a manner as is necessary for our well-being' (PHK §52). This is because expressions of the form 'x causes y' map onto 'x means y' (or 'y means x' when the 'effect' is the sign for the 'cause'), and so can serve the same purposes. As the semantic relation is in place to encourage appropriate action, the use of causal expressions, though false, can lead to the same dispositions to act. Hence fictionalism is sustained because we may retain such utterances inasmuch as they serve a complex, non-representational, psychology and practice.

Still, one might query (among other things) whether, ideally, uses of 'x is the cause of y' should be not be retained, but instead replaced by 'x is a sign for y'. So the vulgar talk is pragmatically serviceable, but the kind of scientific understanding that

Berkeley seeks will regard such relations as properly semantic. This strikes me as correct, but incomplete. For, consider this passage from the *Manuscript Introduction*:

> ...what is it, I prey, to understand perfectly, but only to understand all that is meant by the person who speaks? Which very oft is nothing more than to excite in...the Hearer certain Emotions without any thought of those Ideas so much talk'd of and so little understood (*MI* 43)

This passage raises the following possibility. Whilst it is quite true that the relevant relation between some x and y is correctly represented as 'x means y', and incorrectly represented as 'x causes y', a thinker's *perfect* understanding can consist simply in being appropriately disposed to act in line with the intentions of the speaker. So one might simply be indifferent as to whether that behavioural stance is mediated by the thinker's grasp that 'x means y' or the mistaken assumption that 'x causes y', for one will be in an equivalent state of 'perfect understanding'.

Let us sum up. Berkeley's remarks in the Introduction to the PHK do not entail non-factualism. However, in the context of his discussion of natural causation discourse, they can be usefully viewed as a nascent form of fictionalism. On this view, the semantic content of the relevant discourse is representational, but our acceptance of it and ends served by assertions within it, need not be seen as having a representational function. This is a combination of a non-cognitive psychology with factualist semantics. The relevant form of fictionalism is 'revolutionary' in that it is a recommendation about the use of the discourse in light of the fact that as it stands the discourse implies error. The recommendation, however, is sustained by the fact causal statements both map onto facts about God's linguistic intentions, and the relevant behavioural stances that those semantic relations are intended to enjoin.

3.3 III

As with Berkeley, non-cognitivism is often connected to Hume in the context of his moral theory. Hume scholars in the middle of the twentieth century strenuously argued against readings of Hume as an emotivist,[8] but emotivism is merely one (crude) example of non-cognitivism. More sophisticated non-cognitive readings than emotivist ones are still live options.[9] Our concern, however, is with Hume's position on the causal relation. Notoriously, matters are complex and there are many different interpretative options available.[10] Let us begin with a sketch of the relatively uncontroversial aspects of his account.

According to Hume, for any relation that we deem causal, the cause must be temporally prior to the effect, and that the *relata* must instantiate constant conjunction,

[8] See e.g. Árdal (1966).

[9] E.g. Bricke (1996).

[10] For my own position, see Kail (2007a).

or regularity.[11] Experience of objects standing in this relation produces a new impression, namely an impression of necessary connection, which then 'spread[s] itself on external objects' (T1.4.15: SBN 167).[12] Effectively, then, there is an identifiable input of constantly connected ordered pairs of objects (causation as a 'philosophical relation') that yields a certain subjective response – an impression of necessary connection- which, when combined with the 'objective' side, constitutes causation as a 'natural relation'.

Beyond this, scholarly agreement breaks down. Some see Hume offering a reductive definition of 'cause', and others as offering an account of what is knowable with respect to causation, allowing that there are genuine powers of which we are ignorant. Leaving these options aside, let us consider a third, namely 'projectivism', which is a sophisticated non-cognivitism.[13] In rough outline, projectivism holds that for Hume a) conceptually speaking, the causal relation cannot be reduced to mere regular succession but b) that this is not the same as being committed, ontologically speaking, to genuine powers that outrun mere regularities and c) the absence of such powers does not make causal talk in error because the idea of necessity was never in the business of representing causal powers. Projectivism proposes that Hume's idea of necessary connection should be read as a non-cognitive response to regularities that is 'projected' back onto those regularities. The response accounts for the intuition that, conceptually speaking, we commit ourselves to more that simply 'A always follows B' in claiming that 'A causes B'. We commit ourselves, at a first approximation, to 'A brings about B' or, stronger, 'given A, B *must* occur'. But rather than seeing such a commitment as a fully-fledged representation of a distinct class or modal facts or properties, it is to be interpreted as a *non-representational* response to (perceived) regularities.[14] Its being non-representational is what determines its status as a non-cognitive response. So we get a non-reductive view of causal concepts without any metaphysical commitment to a class of modal properties or facts outrunning regularities.

But what is this non-cognitive response? Commentators tend to take a rather dim view of Hume's account of the 'customary transition' that he identifies as the impression of necessary connection.[15] They argue that there is no genuine account to be found in Hume's writings. However, the account is not as empty as it may seem. In order to understand the account we need first to consider the reasons Hume adduces for his claim that we do not have an impression of genuine necessary connection 'drawn from the objects'. His leading strategy is to consider what cognitive

---

[11] Contiguity, which figures in two 'definitions' of 'cause' in T.1.3.14 is dropped not only in the first *Enquiry* but even within the body of the *Treatise*.

[12] References to Hume's *Treatise* follow the convention of book, part, section, and paragraph number, followed by a page number to the Selby-Bigge/Nidditch editions.

[13] See Blackburn (1984) and (2007); Beebee (2006) and Coventry (2006).

[14] I have said 'perceived' here in order not to foreclose on the possibility that projectivism is in fact compatible with a certain kind of realism about causal powers. For this idea see Craig (2007) and my (2007a).

[15] See e.g. Stroud (1977).

consequences a genuine impression of necessity would yield, and argue that no impressions could yield those consequences. He argues that a genuine impression of necessity would be such as to a) allow the thinker to predict what effect such and such a cause must have prior to the manifestation of that effect in experience and b) render it impossible to conceive a given cause without its effect. But since a) we cannot simply 'read off' what effect any putative cause must have and b) we can always conceive some cause being followed by some other effect, we have no genuine impression of necessity.

Again, this remarkably simple argument masks many complications that we cannot here investigate. The point, as we said, of introducing Hume's negative strategy is to further our understanding of his positive account of the idea of necessity. For the impression of necessary connexion – the 'determination of the mind' – has the effect of mimicking what a genuine impression of necessity is supposed to enable its possessor to do. The impression of necessity we do have is the result of frequent experience of A following B, which consists in a certain change in the phenomenology of causal inference. Rather than making the transition from A to B guided by an explicit grasp that B has followed A in the past, one simply moves *immediately* to the thought or idea of B when A is present. This phenomenology mimics that of moving from A to B in virtue of a grasp of A's being necessarily connected to B. Second, and perhaps more importantly, a genuine grasp of necessity is supposed to render it impossible to conceive some cause without its effect. Now, the customary transition mimics this by substituting a notion of psychological inseparability for conceptual impossibility. Thus Hume writes:

> 'Tis natural for men, in their common and careless way of thinking, to imagine they perceive a connexion betwixt such objects as they have constantly found united together; and because custom has render'd it difficult to *separate* the ideas, they are apt to fancy such a separation to be in itself impossible and absurd (T 1.4.3.9; SBN 223, my emphasis).

So we can see that the idea of necessary connection effects a functional change in the psychology of the thinker, a change that does not owe itself to detecting genuine powers. The impression is not a 'copy' of genuine power and so in that sense it is a non-representational state. Nevertheless, the impression has an effect on the inferential phenomenology of the thinker, and explains, in a non-cognitive fashion, why there is some conceptual difference between 'B following A' and 'A's causing B'. Hence, Hume says, that in the absence of the idea, we might say 'that one object or even has followed another; not that one was produced by the other' (EHU 8.1.5: SBN 82).[16]

[16] References to Hume's *Treatise* follow the convention of book, part, section, and paragraph number, followed by a page number to the Selby-Bigge/Nidditch editions.

3.4 IV

By way of a conclusion, let us now briefly compare Hume's conclusion to the position I attributed to Berkeley. There are two key points of difference. The first that Berkeley's position is attuned to matters regarding linguistic expression in a way that Hume's is not. The *psychology* presented by Berkeley is non-cognitive but it can be expressed in a linguistically descriptive mode. This point is especially interesting, since projectivist readings of Hume see his talk of 'spreading the mind' as a matter (among other things) of linguistic expression. Thus Simon Blackburn writes that we

> ...*project* an attitude or habit or other commitment which is not descriptive onto the world, when we speak and think as though there were a property of things which our sayings describe, which we can reason about, know about, and so on. Projecting is what Hume referred to when he talks of 'gilding and staining all natural objects with the colours borrowed from internal sentiment', or of the mind 'spreading itself on the world' (1984: 170–1).

But, to put matters somewhat briefly and dogmatically,[17] there is just no evidence that Hume's talk of 'spreading the mind' is concerned with the linguistic expression of the idea of necessity. Instead, the metaphor of 'spreading the mind' is invoked to explain a pre-theoretical *perceptual* phenomenology. It is supposed to explain our (illusory) experience of A's bringing about B. So the recognisably non-cognitive component in Hume's philosophy is not matched by a concern with the linguistic expression of this non-cognitive psychology. A second difference is that whilst Berkeley's view is fictionalist, Hume's treatment of the idea of necessity implies non-factualism. Berkeley, recall, did not call into question whether we have a concept of natural causation, but viewed its deployment as yielding falsity. His concern was with the psychology of causal statements and not the status of the concept of natural power. Hume's discussion, on the other hand, focuses on the idea of necessity itself and argues that that idea cannot be construed as a representation of genuine power. So our ordinary practice is instead a matter of a possessing a certain inferential practice rather than a representational one. What is interesting is that twentieth-century non-cognitvisms tended to combine non-factualism, a non-cognitivist psychology, and a concern with linguistic expression in one package. Contemporary philosophy is now taking this package apart, but Berkeley and Hume never combined them.[18]

References

Árdal, P. *Passion and Value in Hume's Treatise* (Edinburgh: Edinburgh University Press, 1966)
Austin, J. L. *How to Do Things With Words* (Oxford: Clarendon Press, 1962)
Beebee, H. *Hume on Causation* (London: Routledge, 2006)

[17]For more sustained discussion, see my (2007a) and my 'Hume, Realism and Quasi-realism', forthcoming.

[18]Thanks to Olivia Bailey, SMS Pearsall, and EMP Kail.

Belfrage, B 'Berkeley's theory of emotive meaning (1708)', *History of European Ideas* 7 (1986), 643–649

Belfrage, B. 'Editor's Commentary' in George Berkeley, *Manuscript Introduction*, ed. Bertil Belfrage (Oxford: Doxa, 1987)

Berkeley, G. *A Treatise Concerning the Principles of Human Knowledge*, ed. J. Dancy (Oxford: Oxford University Press, 1998)

Berkeley, G. *Manuscript Introduction*, ed. Bertil Belfrage (Oxford: Doxa, 1987)

Berman, D. 'Berkeley's Semantic Revolution 19 November 1707–11 January 1708', *History of European Ideas* 7 (1986), 603–607

Berman, D. *George Berkeley: Idealism and the Man* (Oxford: Oxford University Press, 1994)

Blackburn, S. *Spreading the Word: Groundings in the Philosophy of Language* (Oxford: Clarendon Press, 1984)

Blackburn, S. (2007) 'Hume and Thick Connexions' in Richman and Read (eds.) *The New Hume Debate* 2nd ed. (London: Routledge, 2007)

Bricke, J. *Mind and Morality: An Examination of Hume's Moral Psychology* (New York: Oxford University Press, 1996)

Coventry, A. *Hume on Causation: A Quasi-Realist Interpretation* (London: Continuum, 2006)

Craig, E. J. 'Hume on Causation – Projectivist *and* Realist?' in Richman and Read (eds.) *The New Hume Debate* 2nd ed. (London: Routledge, 2007)

Hume, D. *A Treatise Concerning Human Nature*, ed. Norton & Norton (New York: Oxford Unversity Press, 2000a)

Hume, D. *An Enquiry Concerning Human Understanding*, ed. Beauchamp (New York: Oxford University Press, 2000b)

Jakapi, R. 'Emotive Meaning and the Christian Mysteries in Berkeley's *Alciphron*', *British Journal for the History of Philosophy* 10 (2002), 401–411

Jakapi, R. 'Entry 720 of Berkeley's *Philosophical Commentaries* and 'non-cognitive' propositions in Scripture', *Archiv für Geschichte der Philosophie* 85 (2003), 86–90

Kail, P. J. E. *Projection and Realism in Hume's Philosophy* (Oxford: Oxford Unversity Press, 2007a)

Kail, P J E 'Berkeley, the Ends of Language, and *The Principles of Human Knowledge*', *Proceedings of the Aristotelian Society*, 107 (2007b), 265–278

Kail, P J E 'Hume, Realism and Quasi-Realism', in *Recherches sur la Philosophie et le Langage* 26(2009), 117–133

Kalderon, M. (ed.) *Fictionalism in Metaphysics* (Oxford: Oxford Unversity Press, 2005)

Stanley, J. 'Hermeneutic fictionalism' in French & Wettstein (eds.) *Midwest Studies in Philosophy*, Vol. XXV (2001), 36–71

Stroud, B. Hume (London: Routledge, 1977)

Williford, K. 'Berkeley's Theory of Operative Language in the Manuscript Introduction', *British Journal for the History of Philosophy* 11 (2003), 271–301

Part II
Neglected Works and Aspects of Berkeley's Thought

Chapter 4
Berkeley and His Contemporaries: The Question of Mathematical Formalism

Claire Schwartz

4.1 Introduction

Berkeley's critique of the calculus is a well-known topic, as are his attempts to build a brand-new geometry based on sensible minima, but the notion of a Berkeleian mathematical philosophy has hardly been examined. Some recent works have nevertheless tried to analyze what this philosophy could be.[1] Most of these interpretations differ in many respects, but they seem to agree on a few points: particularly on the necessity to consider this philosophy – whether it concerns geometry or algebra – as a differentiated one.[2] More recently, some have emphasized the alleged formalism of Berkeley's thought concerning arithmetic and algebra. Berkeley would thus become representative of strong modernity, anticipating the essence of mathematical formalism and could appear as a major figure in the history of mathematical thought.[3]

We feel that the alleged Berkeleian formalism deserves to be more thoroughly discussed, and qualified. We aim to do this first by attempting to specify the notion of mathematical formalism in order to see how it can be applied to Berkeley, and more broadly, to the mathematics of this period. We will then point out the difficulties that arise when considering Berkeley as a formalist, and, even more, as a precursor to formalism, through two kinds of arguments. Firstly, a comparative and external approach should reveal how these elements of formalism seem to be shared by most of the great philosophers and scientists of his time, who can sometimes be seen as much stronger formalists than Berkeley, as to the new algebraic

---

[1] Baun (1972); Brook (1973); Pycior (1987); Jesseph (1993); Sherry (1993).

[2] Sherry calls it a "porte-manteaux" philosophy of mathematics.

[3] Jesseph, 119, presents Berkeley as "the originator of formalism", although he clearly specifies the limits of such a claim. Baum, Brook and Sherry see some elements of formalism in Berkeley's thought, without considering him as a precursor in this domain.

C. Schwartz (✉)
Université Lyon 3, 97 avenue Philippe Auguste, 75011, Paris, France
e-mail: c.schwartz@freesurf.fr

S. Parigi (ed.), *George Berkeley: Religion and Science in the Age of Enlightenment*, International Archives of the History of Ideas/Archives internationales d'histoire des idées 201, DOI 10.1007/978-90-481-9243-4_4,

operations. Secondly, we will try to show the conflict resulting from the coexistence of formalist and instrumentalist elements in his mathematical thought. We will specify what we mean by instrumentalism, and how, in contrast with a formalist theory, it imposes a denotation to any formal system.

4.2 How to Understand Formalism

4.2.1 Some Features of the Modern Concept of Formalism

The concept of mathematical formalism is largely linked to the constitution of non-Euclidean geometries and the consequent question of the meaning and denotation of mathematical symbols. They are understood as elements of a system whose validity relies upon the consistency and application of certain stated rules. This system may well apply to meanings or real things and relations. The system is then said to be interpreted.

In this sense, formalism implies that:

- The immediate objects of mathematics are signs and operations on mathematical signs.
- Mathematical signs can be understood only in relation to their place in a system structured by some stated rules.
- The interpretation of formal systems does not depend on an intrinsic reference they may denote.

The non-Euclidean geometries would provide an obvious example of formal systems where signs get their meaning out of their integration into a consistent system, and not by their application to perceived extension. Therefore, they can be interpreted in order to denote space in a different physical system.

There are indeed some convincing arguments to claim that Berkeley had understood the formalism of mathematics, that is to say, that they function like formal systems. D. Jesseph, who is inclined to draw this conclusion, bases his demonstration on the rejection of general abstract ideas, and what he calls Berkeley's nominalism.[4] We can also refer to his theory of signification, and in particular to some passages of *Alciphron*, in order to reinforce this claim. Let us examine the elements which could justify the application of the category of formalism to Berkeley's mathematical thought.

[4] Jesseph, op. cit. His reading of Berkeley's mathematical thought is also backed up by the account of Berkeley's youthful texts, in particular *Arithmetica absque Algebra aut Euclide demonstrata*.

4.2.2 *The Formalist Elements of Berkeley's Mathematical Thought*

It is true that Berkeley explicitly claims that:

- The immediate objects of mathematics, at least in the case of arithmetic and algebra, are signs.
- Mathematical signs do not necessarily have a meaning, that is, stand for clear ideas.
- Their function is not to represent, but rather to operate.

These elements are most clearly expressed in *Alciphron* when Euphranor criticizes Alciphron's views, which could more or less be considered as intuitionist.

Euphranor questions the premise according to which words, and more generally signs, stand for distinct ideas. Euphranor points out different kinds of relations between signs and ideas. He illustrates the first with an analogy: just as card players do not always have in mind the money the "counters" stand for when they see and manipulate them, words can be used without their corresponding ideas having to be present in the mind.[5]

Signs are convenient abbreviations of complex ideas, or complex combinations of simple ideas, by which we can connect some ideas more quickly. We are not systematically conscious of the meaning of the ideas the signs stand for, but we always have the possibility to return to them. Likewise, the counter is replaced by the money it stands for at the end of the game.

Euphranor actually wants to go further in his demonstration of the true nature of signs. Their function is certainly to signify – but to signify is not only a question of denoting ideas, but also of inducing some acts:

> It seems also to follow that there may be another use of words besides that of marking and suggesting distinct ideas, to wit, the influencing our conduct and actions, which may be done either by forming rules for us to act by, or by raising certain passions, dispositions, and emotions in our minds.[6]

Therefore, a speech, as a collection of words, can turn out to be meaningful even though each word does not stand for a clear and distinct idea. Euphranor points out the need to consider a broader theory of signification. In this case, words are components of a meaningful system but seem to be devoid of a signification of their own. This claim, which is not about mathematics in particular, is clearly favourable to a formalist interpretation of mathematics as a system of signs.

This conception of signification inevitably leads to a theory of truth, and the conditions by which a system of signs can be said and recognized as true or false. It appears that these formalist elements are organically linked to an instrumentalist conception of truth. Arithmetic and algebraic signs, for instance, do not stand for distinct ideas

---

[5] *Alciphron* (hereafter *Alc.*), VII, 5, *The Works of George Berkeley* (hereafter *Works*), ed. Arthur Luce and Thomas Jessop (Edinburgh, Nelson, 1948–1959), III, 291–292.

[6] *Alc.*, VII, 5, *Works*, 292.

because their very function is not to represent, but to operate. Consequently, what allows us to know whether a proposition is true or false is its utility.

In many texts, Berkeley seems to assert that measuring the utility of a proposition is the best way to know whether it is true or false. This criterion applies just as well to practical as to theoretical discourses. It means that the truth of a proposition can be evaluated by its effects. Berkeley explicitly claims and applies this principle in *Alciphron*:

> We do not deny but there was something useful in the old religions of Rome and Greece, and some other Pagan countries. On the contrary, we freely own they produced some good effects on the people. But then these good effects were owing to the truths contained in those false religions: the truer therefore, the more useful. I believe you will find it a hard matter to produce any useful truth, any moral precept, any salutary principle or notion in any Gentile system, either of religion or philosophy, which is not comprehended in the Christian, and either enforced by stronger motives, or supported by better authority, or carried to a higher point of perfection.[7]

The same must be said for scientific propositions, and that is what Berkeley wants to demonstrate in the Seventh Dialogue of *Alciphron*. Considering that we do not know what force may be, other than by its effects, Berkeley implies that the utility is also what makes us recognize the truth of a physical proposition.[8]

Mathematics are not an exception to this instrumentalist conception of truth. Let us see how Berkeley presents arithmetic and algebra in this same *Dialogue*. Numbers, as objects of arithmetic, are signs standing for things, and then other signs, in arithmetic and algebra, are made up in order to suitably signify their relations and infinite combinations to eventually denote things in the most practical way.[9] Arithmetic and algebra are nothing more than tools: their immediate objects are not ideas, but rather signs by which the mind can relate to things and then act upon them. The case of geometry is slightly different: its objects are not just signs conventionally determined for their efficiency, since they are ideas of perceived extension. Therefore, geometry is useful, as any true discourse, but its immediate objects are not conventional signs and stated rules. Consequently, the formalist interpretation of Berkeley's philosophy of mathematics usually concerns arithmetic and algebra, putting geometry aside.[10] To conclude, what stands out is that, for Berkeley:

- The immediate object of arithmetic and algebra are signs and operations on signs.
- Mathematical signs do not have any intrinsic meaning (except in geometry), that is, they do not stand for ideas.
- As a whole, however, these signs have a meaning, since they usefully refer to things.

---

[7] *Alc.*, V, 10, *Works*, 183; for moral propositions: I, 16, *Works*, 60.

[8] *Alc.*, VII, 7, *Works*, 295.

[9] *Alc.*, VII, 12, *Works*, 304–305; *Principles of human knowledge* (hereafter *PHK*), I, 121, *Works* II, 96–97.

[10] Cf. Brook; Pycior, who considers a tripartite division between geometry/arithmetic and algebra/analysis; Jesseph; MacLeod (2005). In his paper, MacLeod presents some arguments that are very close to ours, in order to question Berkeley's so-called formalism, although this author does not compare Berkeley with some of his contemporaries.

We could consider these elements as heralding mathematical formalism, but we think that this interpretation, albeit attractive, has to be moderated by two kinds of argument. A comparative approach may show how some elements of formalism should be applied, though differently, to the leading philosophers of his time. On the other hand, we can detect some incompatibilities inside Berkeley's philosophy of mathematics between the formalist interpretation of symbolism and his instrumentalist theory of truth.

4.3 The Limits of the Formalist Interpretation

4.3.1 Some Interpretations of Algebraic Symbolism: Descartes, Malebranche, Leibniz

Let us go back to the "counter theory". We said that Berkeley actually goes further in his analysis of symbolism. R. Baum calls it "the weak interpretation" of Berkeley's mathematical formalism.[11] This first level of formalism – rather neglected by Berkeley himself – according to which the significations of signs are put between brackets during the process of reasoning, is analyzed by most of his great contemporaries. Now this sometimes leads to a stronger mathematical formalism based on a subsequent examination of symbolism. On this subject, H. Pycior has remarkably shown how Berkeley fits into the field of mathematics of his time and place,[12] but we would like to turn to a more continental tradition in which we can find new elements in order to discuss the notion of mathematical formalism. We will first focus on a philosopher who had a direct influence on Berkeley: Nicolas Malebranche.

4.3.2 Berkeley and Malebranche

We know that Berkeley read Malebranche's *Recherche de la vérité* carefully.[13] It is thus not surprising to notice some similarities between Berkeley's claims on the nature of arithmetic and algebra and what Malebranche writes about it in *Recherche*'s Book VI, which deals with mathematics. In addition, Malebranche's views are worth considering because he is halfway between the two great mathematical geniuses of his time: Descartes and Leibniz. The mathematical aspects of *Recherche de la vérité* are both based on Descartes' *Regulae ad directionem ingenii*

[11] Baum, 121.

[12] Pycior, 265–286.

[13] As to the relationships between Berkeley and Malebranche, and Berkeley's Malebranchean readings, cf. Luce (1934), especially chapter I.

and on the success of the calculus, which Malebranche discovered through the intermediation of L'Hospital and John Bernoulli.

The comparison we would like to draw now between Malebranche and Berkeley is then relevant in two ways: because Berkeley read Malebranche, and because a closer examination of Malebranche's work on mathematics can help us to measure the originality of the Berkeleian philosophy of mathematics.

4.3.3 Arithmetic and Algebra in the *Recherche de la vérité*

Malebranche is inspired both by Descartes and Leibniz in his analysis of algebraic symbolism; therefore, he enables us to distinguish and identify the different functions ascribed to algebraic reasoning and symbolism during the period in question.

Malebranche separately analyzes geometry on one hand, and arithmetic and algebra on the other, within the context of a meditation on the concept of method. He interprets algebra as an extension of arithmetic, as Berkeley would later do. The utility of geometry is limited to its power of making us attentive, through the use of imagination to suggest exact relations. This idea comes directly from Descartes' *Regulae*.[14] Arithmetic and algebra refer to another kind of utility, as Malebranche claims at the end of the chapter dedicated to geometry:

> Nor have I discussed arithmetic or algebra, because the numbers and letters of the alphabet used in these sciences are useful, not so much for increasing the mind's attention, as for extending its scope, as we shall explain in the following chapter.[15]

Arithmetic and algebra, to some extent, make us attentive through the medium of the signs they are made of, but much less than geometry. The connections between ideas and the traces in the brain are not actually the same concerning geometric, arithmetical and algebraic ideas. In the case of geometry, these connections are "natural", says Malebranche.[16] They do not depend on our will, but result from "the constant and immutable will of the Creator", that is, of nature. For instance, it is not up to us whether we think of a tree when we see a tree, or whether we think of a square when we see a square.[17]

---

[14] *Cf. Regulae* 14, *Oeuvres de Descartes*, ed. Charles Adam and Paul Tannery (hereafter AT), (Paris: Vrin, 1897–1910/1972), X, 438–452.

[15] *The Search after Truth* (hereafter *The Search*), trans. Thomas Lennon and Paul Olscamp (Ohio State University Press, 1980): 430. *Oeuvres complètes de Malebranche*, ed. G. Rodis-Lewis, A. Robinet and others (hereafter *O.C*) (Paris: Vrin, 1958–1984), II, 280.

[16] As to the distinctions of the different kinds of links between ideas and cerebral traces, see II, I, V, *O.C.*, I, 216–22.

[17] "We cannot doubt, for example, that all men have the idea of a square, when they see a square, because this connection is natural", *The Search*, 103; *O.C.*, 219. It is different with the word "square", instituted by the will of men.

The link is of a different kind in the case of arithmetic and algebraic signs. They result from human institution; to signify the surface of a square of side a by aa, for instance, is stated by mathematicians. The notation of numbers has a geography and a history; whether time and use eventually strengthened this link is still arbitrary. This arbitrary nature of the relation between sign and signification is also claimed by Berkeley, but it appears that it is not necessarily a consequence of a nominalist position. For Malebranche, indeed, signs do no stand for things, but rather for ideas, or more exactly, for relations between ideas. The two authors also share the opinion that the nature of the criterion on which symbols are chosen has to be pragmatic: as far as possible, says Malebranche, let us work on signs related to naturally instituted traces in the brain.[18] That will facilitate attention and understanding. Non-natural signs should be used if and only if they can help us to resolve some complex questions. Thus, mathematical symbolism has to be chosen on a pragmatic basis; it does not imply an instrumentalist conception of mathematical truth in any way, and its objects – numbers and indeterminate magnitude – are not ideas but signs.[19]

Malebranche thus draws the conclusion that geometry is easier than arithmetic and algebra, whose real utility is not to help sustain attention. What they are made for is "to increase the mind's scope and capacity".[20]

Geometry preserves the mind's attention, while arithmetic and algebra "skilfully conserve the mind's capacity". This is an original division of labor and there is no trace of it to be found in particular in the Cartesian *Regulae*. Descartes, however, had already analyzed the heuristic function of symbolism. Rule 16 shows how algebra enables us to make up for the lapses in our memory, but the main strength of this science is "*difficultatis terminos exhibere*".[21] In this text, Descartes clearly identifies two functions of mathematical symbolism:

- To spare memory and to focus attention on reasoning and not on recollection
- To make the relations between the elements of a problem appear

For instance, algebraic symbolism makes the complexity of a problem appear immediately, by showing whether it is an equation of first, second, or third degree, which magnitudes are of the same dimension, etc.

These functions are also present in Malebranche's account of arithmetic and algebra. They enable us to recognize truths, that is, relations between ideas, since Malebranche defines truth as nothing but a real relation whether of equality or inequality.[22] But he has to prove that arithmetic and algebra are the most efficient and direct means to "skilfully conserve the mind's capacity" and that formalism is

[18] "But as far as possible, one must use accepted terms or terms whose ordinary signification is not so far removed from what one is claiming to introduce, and this is not always observed in mathematics.": *The Search*, 105; *O.C.*, I, 221.

[19] That is what Berlioz (2000), 146, seems to claim.

[20] *The Search*, 431.

[21] AT X, 455.

[22] *O.C.*, II, 286, *The Search*, 433.

helpful in order to reach that goal. According to Malebranche, arithmetic has two features which makes this possible. First, the ingenious notation based on a few figures allows the expression of all numbers according to their relation to unit, and all their relations by the combinations of very simple symbols. For instance, $4321 = 4.1000 + 3.100 + 2.10 + 1$ is a sum of four magnitudes constituted by the product of two magnitudes: a figure and a multiple of ten. The notation spares us from having to think about all of these relations successively, and allows us instead to operate directly on the quantity chosen.

The second strength of arithmetic is to reveal truths on the composition of numbers. Let us take the example of 8^2 and how to calculate it. We can factorize this product, by the laws of analysis[23]: $8.8 = (5 + 3)^2 = 5^2 + 2.(5.3) + 3^2$. The factorization enables us to reduce the calculation of a square number to the squares of inferior numbers, more easily known. Malebranche then outlines these arithmetical processes:

> Thus, arithmetic provides the means of expressing all the simple and complex relations possible between magnitudes. It then shows us how to perform the calculations to deduce these relations from one another, and to discover the relations of magnitudes that might be useful by means of those already known; and it shows us how to do this with skill, with clarity, and with a remarkable exploitation of the mind's meager capacity.[24]

Algebra, as the use of literal notations of mathematical relations in which letters can be substituted by any number, generalizes the arithmetical results. For instance, the particular truth we mentioned above corresponds to the general algebraic formula: $(a + b)^2 = a^2 + 2ab + b^2$.

Can we now consider this Malebranchean account of mathematical formalism as nothing but a "counter theory", or a weak interpretation of formalism? It is indeed a question of using some symbols whose meaning is put between brackets. For instance, once we state that *a* is the symbol for the side of a square, we do not think of its signification when it is present in a system of equations where other letters can stand for other geometric magnitudes. We just work on the combinations of signs, according to algebraic rules. But at the end of the calculation, their meaning is restored, as long as we use algebra in order to resolve geometric problems. Furthermore, letters can arbitrarily stand for any magnitude, as the counter can for any amount of money arbitrarily stated. This is made possible by the algebraic process of Descartes' *Géométrie*, by which any magnitude can be chosen as an arbitrary unit.[25]

Descartes and Malebranche, however, also emphasize another function of mathematical formalism: its heuristic power. They show how the formalization of a problem according to an ingenious symbolism helps us to understand its complexity or to reach some results faster and easier. It makes an intrinsic dynamism of symbolism

---

[23] Malebranche calls analysis what we would call classical algebra, that is, the theory of equations: *O.C.*, II, 293.

[24] *The Search*, 435, *O.C.*, II, 291.

[25] AT VI, 449.

appear: the manipulation of symbols helps to discover new results, that is, relations between magnitudes. If then we have to ascribe to Malebranche and Descartes the weak interpretation of formalism, because the signification of signs is only suspended during the process of reasoning, it goes further than the "counter theory". It could be considered as a formalist element of their philosophy of mathematics which Berkeley did not analyze. Even if he did, he was certainly not the first.

Furthermore, it is not clear that algebra, for Descartes, is in the service of geometry. His *mathesis universalis*, though later abandoned, is based on a general theory of magnitude, or science of proportions. In his *Géométrie*, it is more geometry that can be considered as being in the service of algebra.[26] If mathematical symbols have a meaning, it is a very general one, that is, of magnitude, or more exactly, of finite ratios between magnitudes. This is why we can freely choose symbols and state their meaning since they do not stand in essence for any determined magnitude. This is also the case for Malebranche, who defines truth as a relation between ideas, and the object of mathematics as ratios or relations between magnitudes. Symbols, or combinations of symbols, stand for general relations, rather than specific ideas.

It is obvious that this kind of formalism is incompatible with Berkeley's rejection of general abstract ideas. Berkeley does not actually deny the existence of general ideas, or more exactly, of *general meanings*:

> An idea, which considered in itself is particular, becomes general in its meaning by being made to represent or stand for all other particular ideas of the same sort as itself. Suppose for example that a geometrician, proving the validity of a procedure for cutting a line in two equal parts, draws a black line one inch long. As used in this geometrical proof, this particular line is general in its significance because it is used to represent all particular lines, so that what is proved regarding it is proved to hold for all lines. And just as that particular *line* becomes general by being used as a sign, so the *word* 'line' – which in itself is particular – is used as a sign with a general meaning. The line is general because it is the sign not of an abstract or general line but of all particular straight lines that could exist, and the word is general for the same reason – namely that it stands equally well for each and every particular line.[27]

Berkeley tries to match his nominalism with the generality of mathematical demonstration. It could work for extension: a particular line could stand for all lines, and be considered as a sign with a general meaning. But which particular idea could stand, not for anything extended, or counted, but for all magnitudes, or quantities? Is not magnitude a general idea in itself?

It is nonetheless true that, while for Descartes and Malebranche symbols stand for magnitudes or relations between real or possible magnitudes, Berkeley seems to go further. He claims that meaning does not have to be necessarily ascribed to signs, considered separately, but to a whole system of signs, considered as meaningful according to its effects. There was one mathematical question, at that time, which obviously implied such formalist considerations: the calculus and the status of infinitesimal quantities.

---

[26] On the algebraic signification of the Cartesian *Geometry*, see Boutroux (1900); Boyer (1959); Jullien (1996).

[27] *PHK*, Introduction, *Works* II, §12.

4.3.4 *Leibniz and the Problem of Infinitesimal Quantities: A Stronger Interpretation of Formalism?*

Let us now compare Berkeley and Leibniz, the latter who could be considered a strong formalist, especially in his discovery of the calculus. The formalist elements of Leibniz's thought are also related to his plan for a universal characteristic, which has two meanings. In the first case, it constitutes a universal and ideal language, composed of characters expressing the real content of concepts. In the second case, it concerns more the possibility to formally express the logical relations between concepts, whose content is ignored and with which we substitute symbols. In this context, Leibniz does not insist on what characters express.[28] The use of such a language would make us reason mechanically while calculating:

> Mais ce serait le moindre de ses avantages, car cette même écriture serait une espèce d'Algèbre générale et donnerait moyen de raisonner en calculant, de sorte qu'au lieu de disputer, on pourrait dire: comptons. Et il se trouverait que les erreurs de raisonnement ne seraient que des erreurs de calcul qu'on découvrirait par des épreuves comme dans l'arithmétique.[29]

In this case, Leibniz calls it "spécieuse générale":

> J'oserais ajouter une chose, c'est que si j'avais été moins distrait, ou si j'étais plus jeune, ou assisté par de jeunes gens bien disposés, j'espèrerais donner une manière de Spécieuse Générale, où toutes les vérités de raison seraient réduites à une façon de calcul. Ce pourrait être en même temps une manière de langue ou d'écriture universelle, mais infiniment différente de toutes celles qu'on a projetées jusqu'ici, car les caractères et les paroles mêmes y dirigeraient la raison, et les erreurs (exceptées celles de fait) n'y seraient que des erreurs de calcul.[30]

What is put forward is the automatic calculation that the use of this language implies, and that is made possible by the judicious choice of symbols. Such a language guarantees correct reasoning, and helps the mind to establish the right connections. We do not consider *what* we reason *on.* It is actually close to what Malebranche says about the fecundity of arithmetical and algebraic operations.

Do we therefore have to reduce the functions of the Leibnizian "spécieuse" to the Malebranchean account of mathematical formalism? This would be to ignore what Leibniz himself considers as one of the greatest successes of his characteristic, which Malebranche knew but Descartes could not accept: the calculus. In this context, Leibniz introduces the interesting concept of symbolic, or blind thought. It first appears in the categorization of the different sorts of ideas established by Leibniz in *Meditationes de cognitione, veritate et ideis.* When the analysis of our ideas is "very long", we have to use characters which stand for them, and prevent

[28] Heinekamp (1975) stands up for the first interpretation and Russell (1900) for the second. See also Rutherford (1995).

[29] G. W. Leibniz, *Die Philosophischen Schriften,* ed. Carl I. Gerhardt (hereafter GP), (Berlin: Meidmanniche Buchhandlung, 1875–1890), VII, 26.

[30] *À Remond,* 10 novembre 1714, GP III, 605.

our having to think of the actual content of such ideas. *A fortiori*, we also have to substitute a character to the analysis of ideas when the analysis has to be infinite.[31] In this case, blind thought is not only useful or efficient, but necessary. Let us see how it is applied to the signs of differential and integral calculus. The sign $\int$ falls necessarily within blind thought, since the mind can obviously not adequately embrace all the infinitesimal summations implied by this operation. This sign, then, signifies an operation which can never be represented to the mind as a construction made of a finite number of simple ideas. Furthermore, it is not clear what we operate on in this case. Leibniz actually has some trouble when he is asked to define the differentials. When he introduces them for the first time in his *Nova Methodus*, he does not define them clearly, but gives the adequate rules by which we can operate on them.[32] This would be the strong version of mathematical formalism: we do not know what the signs stand for, since they get their meaning from their integration to a correct formalism, with its rules of composition. In this case, the manipulation of signs produces within itself a signification which cannot be perceived independently of this formal operation. It is just as if, for Leibniz, the infinitesimals were validated by their effects, that is, the efficiency of reasoning and the exactitude of the results to which they lead us.

Is this not exactly what Berkeley claimed concerning mathematical symbolism, and more generally, about truth? Did he, then, explicitly assert what Leibniz was doing without clearly stating it? Paradoxically, as we know, Berkeley rejected the infinitesimals. This is not exactly the sign of a genuine formalist thinker. One of the reasons for this rejection is that Berkeley does not consider magnitude as the unique object of mathematics, whether it be continuous or discrete. Geometry is considered as denotative, algebra as operative. Thus he reintroduces the divorce between algebra and geometry, to which the Cartesian *Geometry* put an end. Leibniz, with some non-Cartesian concepts, extended this Cartesian revolution to new curves.

This is how we are actually led to the internal problem of the Berkeleian account of mathematics and mathematical reference. If Descartes, Malebranche and even Leibniz can be considered as less formalist than Berkeley, since they usually claim that signs must have a signification (that is, they stand for ideas as relations between quantities), they could admit some formal systems expressing all the possible quantities without worrying about their application to things. It is, in any case, guaranteed that these systems will find an interpretation, in particular in geometry (being a science of continuous magnitude). This is an aspect of the elements of a formalist philosophy of mathematics that Berkeley could not have conceived.

---

[31] In the *Meditationes*, Leibniz takes the Cartesian example of a chiliogon, whose analysis is actually very long, but not infinite.

[32] *Nova Methodus pro maximis et minimis, Acta Eruditorum*, October 1684, G. W. Leibniz, *Die Mathematische Schriften*, ed. by Carl I. Gerhardt (hereafter GM), (Berlin, 1849–63, reed. Hildesheim, New York: Olms Verlag, 1971), V, 220–226. About the Leibnizian discussion on the infinitesimals, see: Belaval (1960); Parmentier (1989); Burbage and Chouchan (1993); Mancosu (1996); Eberhard (1999).

4.3.5 The Internal Constraints of Berkeley's Account of Mathematics

There is a formalist element in Berkeley's philosophy of mathematics which cannot be denied: algebraic signs do not have any intrinsic meaning, that is, they do not stand for clear ideas. They become meaningful as part of a set of signs, defined by some stated rules of manipulation.

On the other hand, his philosophy tends to lead him away from a conception of mathematics as a formal system. Arithmetical and algebraic relations, in his sense, do not describe general relations between quantities (true in all possible worlds, susceptible to indefinite interpretations), but they describe the relations between things, that is, perceived things. Arithmetic and algebra are tools that enable us to act upon things, since they describe their relations correctly. Their utility, as we have seen, is the proof of their truth. The non-denotating and operating aspect of these two disciplines is not related to free operations made on symbols, but to determined and useful actions made on existing things.

This is incompatible with authentic formalism, and takes us back, to some extent, to an "Euclidean" conception of mathematics, not only for geometry, which describes perceived extension, but also for arithmetic and algebra – which relate to, if not represent, things.

So there is obviously a conflict between the formalist interpretation of Berkeley's philosophy of mathematics and the instrumentalist elements of his theory of verification. His instrumentalism imposes a denotation to mathematical signs, which is absolutely contrary to the very concept of modern formalism.[33] In other words, it is just as if Berkeley broadened the field of meaningful terms while at the same time restricting the denotation of formal systems. If mathematics were a free game of symbols determined by some stated rules, the symbols just could not be in essence a tool to refer to existing things. On this account, Berkeley appears much more as a nominalist and an instrumentalist than as a pure formalist.

4.4 Conclusion

We can now measure the difficulties of applying the concept of formalism to the mathematical philosophies of this period, and more specifically to Berkeley's, which is considered a precursor to such a conception of mathematics. None of the philosophers we mentioned fully anticipated what can be considered as the main characteristics of formalism. It looks as if they each tried to integrate the new results and operations of algebraic analysis and the way they understood them to their respective philosophy, and in particular to their ontology of mathematical objects. Descartes, Malebranche and Leibniz could be considered as more advanced with

[33] McLeod also discusses this problem in his paper.

the notion of a free constitution of formal systems without any specific interpretation, since these formal systems refer to the general idea of magnitude or general ideas between magnitudes, which are true in all possible worlds. However, contrary to Berkeley, they did not explicitly claim that signs, considered separately and through the denotation of implicit operations rather than ideas, could simply have no meaning. Berkeley's claim is indeed very innovative, but it needs to be considered within the scope of his philosophy in order to qualify, as we have seen, any retrospective reading of his mathematical thought.

References

Baum, Robert. "The Instrumentalist and Formalist Elements of Berkeley's Philosophy of Mathematics", *Studies in History and Philosophy of Science* 3 (1972): 119–134.

Belaval, Yvon. *Leibniz critique de Descartes.* Paris: Gallimard, 1960.

Berkeley, George. Alciphron, or *The Minute Philosopher*, in *The Works*, edited by Arthur A. Luce and Thomas E. Jessop. Edinburgh: Thomas Nelson and Sons, 1948–1959, vol. III.

Berlioz, Dominique. *Berkeley: un nominaliste réaliste.* Paris: Vrin, 2000.

Boutroux, Pierre. *L'imagination et les mathématiques selon Descartes.* Paris: Alcan, 1900.

Boyer, Carl B. "Descartes and the Geometrization of Algebra", *The American Mathematical Monthly* 66, n. 5 (1959): 390–393.

Brook, Richard. *Berkeley's Philosophy of Science.* The Hague: Martinus Nijhoff, 1973.

Burbage Franck, and Chouchan Nathalie. *Leibniz et l'infini.* Paris: Presses universitaires de France, 1993.

Descartes, Réné. *Regulae ad directionem ingenii*, in *Oeuvres*, edited by Charles Adam and Paul Tannery. Paris: Vrin, 1897–1910/1972, vol. X.

Knobloch Eberhard. "Galileo and Leibniz: Different Approaches to Infinity", *Archive for History of Exact Sciences* 54 (1999): 87–99.

Heinekamp, Albert. "Natürliche Sprache und Allgemeine Charakteristik bei Leibniz", *Studia Leibnitiana Supplementa* 15 (1975): 257–286.

Jesseph, Douglas M. *Berkeley's Philosophy of Mathematics.* Chicago: University of Chicago Press, 1993.

Jullien, Vincent. *La Géométrie de 1637.* Paris: Presses universitaires de France, 1996.

Leibniz, Gottfried Wilhelm. *Die Mathematische Schriften*, ed. by Carl I. Gerhardt. Berlin, 1849–1863, reed. Hildesheim, New York: Georg Olms, 1971. *Die Philosophischen Schriften*, edited by Carl I. Gerhardt. Berlin: Meidmanniche Buchhandlung, 1875–1890.

Luce, Arthur A. *Berkeley and Malebranche.* Oxford : Oxford University Press, 1934.

Mac Leod, Miles. "The Formalist Interpretation of Berkeley's Philosophy of Mathematics". Paper presented at the International Berkeley Society Conference, Tartu, Estonia, 5–8 September, 2005.

Malebranche, Nicolas. *The Search After Truth*, translated by Thomas Lennon and Paul Olscamp. Ohio State University Press, 1980. *Recherche de la vérité, in Oeuvres complètes* edited by Geneviève Rodis-Lewis, A. Robinet and others. Paris: Vrin, 1958–1984, vol. II.

Mancosu, Paolo. *Philosophy of Mathematics and Mathematical Practice in the Seventeenth Century.* Oxford: Oxford University Press, 1996.

Parmentier, Marc. *Naissance du calcul différentiel.* Paris: Vrin, 1989.

Pycior, Helen. "Mathematics and Philosophy: Wallis, Hobbes, Barrow, and Berkeley", *Journal of the History of Ideas* 48 (1987): 265–286.

Russell, Bertrand. *A Critical Exposition of the Philosophy of Leibniz.* Cambridge: Cambridge University Press, 1900.

Rutherford, Donald. "Philosophy and Language in Leibniz", in *The Cambridge Companion to Leibniz,* edited by Nicholas Jolley. Cambridge: Cambridge University Press (1995): 124–176.

Sherry, David. "Don't Take Half The Way: on Berkeley's Mathematical Reasoning", *Studies in History and Philosophy of Science* 24, 2 (1993): 207–225.

Chapter 5
Locke, Berkeley and Hume as Philosophers of Money*

An Apology and Synopsis

George C. Caffentzis

1. *The problematic of the paper: What is it to be a philosopher of money and are Locke, Berkeley and Hume philosophers of money?*

For the last 30 years I have been writing a trilogy on Locke's, Berkeley's, and Hume's philosophies of money. With the publication of *Clipped Coins. Abused Words and Civil Government; John Locke's Philosophy of Money* and *Exciting the Industry of Mankind; George Berkeley's Philosophy of Money* and with the last volume on Hume in preparation, the trilogy is now almost completed.[1] But as I near the end of the project, I realize that I have not been as precise as possible in defining what is a philosopher of money, especially given the fact that there are a number of different plausible definitions in the field. Nor have I made it clear in what particular sense are Locke, Berkeley and Hume philosophers of money while, I would say, Bertrand Russell is not. In Berkeley's case, for example, my lack of clarity is especially egregious, since I wrote a 450-page book on *The Querist*, a supposedly "marginal" text of 80 pages consisting all and only of questions.

Why was my philosophy of money trilogy written as it was? I will address this question (and provide an apology for myself and a synopsis for the reader) by first answering the general question, What is a philosopher of money? and then I will go on to show that Locke, Berkeley and Hume fit the definition.

---

*This paper has been presented at the International Berkeley Conference at the Redwood Library in Newport, Rhode Island, June 28, 2008.

[1] See Caffentzis (1989); Caffentzis (2000); Caffentzis (2001); Caffentzis (2005); Caffentzis (2008).

G.C. Caffentzis (✉)
Department of Philosophy, University of Southern Maine, P.O. Box 9300,
04104-9300 Portland, Maine, USA
e-mail: GCAFFENTZ@aol.com

S. Parigi (ed.), *George Berkeley: Religion and Science in the Age of Enlightenment*,
International Archives of the History of Ideas/Archives internationales d'histoire des idées 201,
DOI 10.1007/978-90-481-9243-4_5, © Springer Science+Business Media B.V. 2010

2. *Two models of what it is to be a philosopher of money: (A) being a philosopher and being a monetary commentator; (B) Simmel's neo-Kantian perspective on being a philosopher of money.*

One easy way to define what it is to be a philosopher of money is as a *conjunction* of being a philosopher *and* being a monetary commentator. A surprising number of philosophers have commented on money, often profoundly and perceptively, from Heraclitus to Heidegger. In fact, it would not be difficult to edit a substantial anthology of philosophers' writings on money from the pre-Socratics to the postmoderns.

Such an anthology would be useful to both philosophy and economics. It would undoubtedly give pride of place to the monetary writings of Locke, Berkeley, and Hume. For these philosophers devoted substantial pamphlets, essays and books on money. Their contributions are not just passing metaphorical use of monetary phenomena as, for example, Kant's use of the example of the "actual" taler in the pocket versus the concept of the taler to make the point that existence is *not* a predicate or in his evocation of commercial life to make the same point: "the attempt to establish the existence of a supreme being by means of the famous ontological argument of Descartes is merely so much labor and effort lost; we can no more extend our stock of [theoretical] insight by mere ideas than a merchant can better his position by adding a few noughts to his cash account."[2]

Although contemporary philosophers are rarely mentioned in the standard histories of economics (e.g., Derrida's work on money and the gift has received little notice in economics journals), important contributions to monetary knowledge (laws, observations, concepts) are often attributed to philosophers of the past like Locke, Berkeley and Hume in these texts.[3] At times, connections are even drawn between their philosophy and their monetary theory. This tradition goes back to the nineteenth century and Marx's writings on the history of political economy, from his *Contributions to a Critique of Political Economy* to *Theories of Surplus Value.* In those texts Marx occasionally referred to Locke's, Berkeley's, and Hume's contributions to the theory of money alongside their philosophical efforts. For example, he wrote: "Very fittingly, it was Bishop Berkeley, the advocate of mystical idealism in English philosophy, who gave the doctrine of the nominal standard of money a theoretical twist."[4] This type of crocheting with philosophy – where the conjunct*ion* between philosophy and monetary theory is taken as a temporary conjunct*ure* – continues to be found in the literature on the history of economics and money in the

[2] Quoted in Shell (1982), 139.

[3] See Derrida (1992) and Blaug *et al.* (1995). Though there is ample recognition of the achievements of Locke, Berkeley and Hume in the realm of monetary theory and policy, it is by no means equally spread. The differential attention paid to Locke, Berkeley, and Hume in the history of economics literature can easily be judged by looking at some standard textbooks in the field. Consider Blaug's (1968). It has 21 index [page] references for Hume, 13 for Locke, and 3 for Berkeley. Spiegel's textbook (Spiegel 1983), which is more oriented to the humanities, shows a similar differential: 48 index references to Locke, 32 for Hume, and none for Berkeley. I discuss this differential in Caffentzis (2000), 419n.

[4] Marx (1970), 78.

twentieth century. For example, Joseph Johnston in his essay, "Locke, Berkeley and Hume as Monetary Theorists," characterizes Locke's attitude toward gold and silver in the following words: "[Locke] regarded the value of money as inseparably associated with the precious metals – a primary quality of theirs, so to speak."[5]

This "so to speak" approach can be found in many other works in the history of economics. But it does not reveal any commitment to unearthing conceptual structures common to a philosopher's metaphysics or epistemology and his/her theory of money. However, this conjunction model has been recently taken to the limit by a postmodernist effort to read philosophy as symbolic production and to argue that all philosophy written in a monetary society is a sort of philosophy of money. This rather expansive approach to the philosophy of money is attractive, since it makes it possible to create a rich intertextual field of literary, philosophical and economics writing as an object of study. Marc Shell, one of the founders of this field, writes of his approach:

> This participation of economic form in literature and philosophy, even in the discourse about truth, is defined neither by what literature or philosophy talk about (sometimes money, sometimes not) nor by why they talk about it (sometimes for money, sometimes not) but rather by the tropic interaction between economic and linguistic symbolization and production. A formal money of the mind informs all discourse and is as unaffected by whether or not the thematic content of a particular work includes money as by whether or not the material content of the ink in which the work may be inscribed includes gold.[6]

Thus, the "tropic interaction" between economics and philosophy that was unleashed by the development of monetary economies in the early modern era affects philosophical writings on *any* subject. The interpreter is then free to create interesting monetary readings of apparently non-monetary texts since, as Shell reveals, "my argument is not that money is talked about in particular works of literature and philosophy (which is certainly the case), but that money talks in and through discourse in general."[7] In that sense, every philosopher in a monetary society is a philosopher of money whether s/he designs to be or not. The occasional conjunction of philosophy and money is in this approach taken as a superficial sign of a deeper universal conjuncture.

The second model for a philosopher of money arises from the ur-text of the field, Georg Simmel's *Philosophy of Money*. This book, published in 1900, not only gave a name to a discipline, but it provided the discipline with a methodological structure rooted in the Kantian tradition (justifying Simmel's place in the canon of Neo-Kantian philosophers).

Simmel follows the structure of the *Critique of Pure Reason* as literally as possible in his *Philosophy of Money*. Kant organized his *Critique of Pure Reason* in two major divisions (reserving the Transcendental Aesthetic as an antecedent to the first division): the transcendental analytic and the transcendental dialectic.

---

[5] Johnston (1970), 84.

[6] Shell (1982), 4.

[7] Shell (1982), 180.

The transcendental analytic was directed to determining the necessary preconditions for the creation of objective experience. Hence Kant's analysis of substance, causality, and simultaneity discovers something like "a new storey beneath" empiricism (to use a Simmelian metaphor out of context).

The transcendental dialectic is based on the critique of the major ideas that have populated philosophy from its beginning: soul, cosmos, and god. In this division Kant attempts to demonstrate via paralogisms, antinomies, and impossibility proofs that any attempt to treat these ideas as objects with fixed properties is doomed to failure. They are "transcendental illusions" even though they are the natural result of the operation of reason. Kant, therefore, sees the proper role of philosophy as dealing with the lower and upper bounds of objective experience and not the details of this experience itself.

Simmel begins *The Philosophy of Money* on the vertical dimension with a nod to Kant: "Every area of research has two boundaries marking the point at which the process of reflections ceases to be exact and takes on a philosophical character."[8] He rephrases this point on the horizontal dimension a bit later on in his exposition: "If there is to be a philosophy of money, then it can only lie on either side of the economic science of money."[9] In fact, he disclaims any confrontation with the discipline of economics (which was just beginning to find its own autonomy in his era) that lies between the philosophical areas: "Not a single line of these investigations is meant to be a statement about economics."[10]

Just as Kant did in his *Critique*, Simmel divides his work into two parts: an "Analytical Part" and a "Synthetic Part." The former "present[s] the pre-conditions that ... give money its meaning and practical position."[11] It definitely evokes Kant's Transcendental Analytic and the deduction of the pre-conditions of objective experience, for Simmel is not interested in the origin and realization of money in history, but in the "mental states, in social relations and in the logical structure of reality and values" that constitute money's preconditions (i.e., what makes money possible). The Synthetic Part takes the place of Kant's Transcendental Dialectic in that it too deals with totalities like "life in general," "the inner life," "culture in general" and "the ultimate values and things of importance in all that is human." It attempts to make them intelligible from the "effectiveness of money."[12]

So, for example, "the fact that two people exchange their products" is not only an economic fact; it is also "the object of philosophical study, which examines its preconditions in non-economic concepts and facts [the Analytical Part] and its consequences for non-economic values and relationships [the Synthetic Part]."[13] Simmel practiced this Neo-Kantian methodology by first analyzing the necessary conditions

[8] Simmel (2000). 53.
[9] Simmel (2000), 54.
[10] Simmel (2000), 54.
[11] Simmel (2000), 54.
[12] Simmel (2000), 54.
[13] Simmel (2000), 55.

for the categories of value, substance and the sequence of purposes to appear and create a monetary universe. He then proceeds to deal with the contradictions and dilemmas created by money for totalities like the self and the style of life.

3. *A critique of these models: the first is too broad, and the second too narrow and the proposal of a third.*

Although very attractive, I found these two models for defining what it is to be a philosopher of money to be problematic.

The first is too broad in scope and invites one to interpret every reference (and, as will be shown below, non-reference) to monetary phenomena in a philosopher's work as a contribution to the philosophy of money. But that is no more defensible than taking a philosopher's reference to the cat on the mat as a contribution to the philosophy of biology. Indeed, given the ubiquity of monetary references, almost every philosopher in history would become a philosopher of money according to this definition.

The postmodern "tropic interaction" approach of Shell and others that extremizes the "conjunction" approach cannot be dismissed so easily. But it requires a methodology that is determined by the text's literary quality (its tropic force and extension) and not by its philosophical character. In a monetary society, as I read Shell, just as every poet is a poet of money and every dramatist is a dramatist of money, so too every philosopher is a philosopher of money.

This might very well be true, i.e., the figurative power of money is so deeply embedded in the collective imaginary of monetary societies that every other symbolic production is either directly or, more likely, indirectly affected by it, philosophical writing included. However, this approach does not provide a distinctive role for philosophical thought and writing. One is a philosopher of money simply because one is a philosopher … it all happens behind everyone's back. But, conversely, in such a night there is no reason to discern any particular philosopher *as* a philosopher of money either. In the very moment that this approach defines the philosophy of money, the field disappears in a fog of metaphor, allusion, and, at best, brilliant association.

The second model – based on Simmel's *Philosophy of Money* – has the opposite problem, that of over-specificity. Not only do we have in Simmel's model a specific Neo-Kantian conception of philosophical analysis – a study of money's preconditions and the dilemmas it poses for associated totalities – but there is also a strict separation between economic facts and philosophical investigations.

This strict separation is problematic both philosophically and historically.

As Quine, Sellars, and Kuhn (among many others) taught us in the last philosophical generation, there is no defensible dichotomy between the analytic and synthetic, between science and philosophy, between fact and meta-fact that Simmel's Neo-Kantianism presupposes. In particular, the science of economics cannot be hermetically kept at a distance from "philosophical contamination" for two reasons: (a) this dichotomy is anachronistic and (b) it does not recognize the fact, even down to our time, that money itself is not a fixed concept defining a definitive object of study, i.e., there are continual revolutions in the theory and practice of money.

The assumption's anachronism can be caught in noting that the very name of the field that we now call "economics" did not exist in its contemporary meaning until the later part of the nineteenth century. "Political Economy" was the disciplinary name that existed immediately before "economics" to refer to the study of (roughly) exchange, production and money while, since Xenophon's time, "oikonomia" largely meant the study of the management of the home-based enterprises. But neither "political economy" nor "oikonomia" are synonymous with "economics." Therefore, any attempt to separate philosophy from "economic" facts and theories would be inoperative for philosophers working before the 1870s and would make a Simmelian definition of a philosopher of money inoperative.

An even more important historical point of criticism is that money itself has undergone many transformations and revolutions both in its theory and practice that have incorporated philosophical concepts and methods. Philosophy has played a role in monetary revolutions as it has in the scientific revolutions of physics and other natural sciences. That is inevitable, since philosophy has specialized in the production of concepts like value, fairness, and equality that the creators of a monetary institution must possess and apply both reflexively and mutually. Dichotomizing philosophy and money (as Simmel's methodology requires) makes these monetary revolutions unintelligible. It puts one in a position similar to a historian of science trying to understand the origin of Einstein's relativity theory without a study of the philosophical debates concerning space and time current in the late nineteenth century. In other words, though money might, as Shell claims, "talk in and through discourse in general," philosophy has at times "talked in and through" money itself and one cannot understand money without attending to this philosophical talk.

This last criticism of a Simmelian definition of what it is to be a philosopher of money provides a good segue to the definition I have been using in my work. I argue that philosophers of money are those philosophers who actually try to "change the (monetary) world." Neither are they just "influenced" (or spoken through) by money nor do they simply "reflect" on the monetary world at an infinite distance. They are neither unconscious and passive ideologists nor totally conscious and detached observers of the monetary universe.

Philosophers of money (including the monetary nihilists like Diogenes) are committed to a monetary program of action. Their philosophy constitutes and subsumes a monetary act. Consequently, one can only understand their philosophy by contextualizing it and explaining it the way every other historical act is explained. That is why establishing whether a philosopher is a philosopher of money requires historical evidence and a historical explanatory structure (a context, a set of collective interests and association, an oppositional force, etc.)

4. *This approach leads one to see the philosophy of money as a strategic science that constructs categories to deal with monetary crises and revolutions. The philosopher's system of concepts is applied to the construction of the theory of money and monetary reality has an impact on their philosophy.*

A corollary of this definition is that a philosopher of money's philosophy (ontology, epistemology, metaphysics, and ethics) must have a strategic character and be deployed in the promotion of a monetary program. Indeed, if there were no evidence of such a deployment, there would be no reason to identify a philosopher as a philosopher of money. It is exactly in the use of philosophical concepts in the development and justification of monetary forms that the work of the philosopher of money is done. It is here that the analytic work of the student (and practitioner) of the philosophy of money reveals itself and proves its explanatory worth. For philosophical argument can change the (monetary) world because of the conceptual character of money itself.

The relationship between a philosopher of money's philosophy and monetary reality is not solely active, however; for a philosopher of money's philosophy must also be affected by money. Again, if there were no evidence of this impact of money on his/her philosophical categories, then s/he would not be a philosopher of money. I am not hypothesizing a universal "money of the mind" here. For there can be philosophers who have commented more or less extensively on money but have not had their philosophies deeply impacted by monetary crises and revolutions.

On the basis of this definition of what is a philosopher of money, not every philosopher who writes about money is a philosopher of money, but a philosopher who does not write about money is, *pace* Shell, *not* a philosopher of money. Thus there is a definite categorical problem when asking "Is X a philosopher of money?" (where X might be Plato, Georg Simmel, John Stuart Mill, St. Augustine, Isaac Newton, Simone De Beauvoir, or Martin Heidegger). Its answer requires both philosophical analysis and historical investigation of the following questions: (a) What is the philosopher's monetary program and strategy? (b) How are his/her philosophical categories deployed in the construction and defense of his/her monetary program? (c) Does money play an important role in his/her philosophy?

Not accidentally, Locke, Berkeley and Hume fit the definition of philosophers of money and in the next part of this paper I will sketch out answers to these questions concerning these philosophers (and provide a synopsis of the trilogy) with the help of the following table:

	Locke	Berkeley	Hume
Texts	*Some Considerations...* (1692); *Further Considerations...* (1696)	*The Querist* (1735–1737)	*Political Discourses* (1752)
Crisis	The recoinage crisis of 1696	The tithe agistment crisis of 1734	The "45er"
Circle	The Shaftsbury Whigs and the "Junto"	The Dublin Society and the Anglican Church of Ireland	The "Enlightenment" intelligentsia of Scotland
Opposition	Clippers, "Coiners," Baggers and the "inflationists"	Irish "natives" and the Anglo-Irish gentry	The Highland rebels, the mercantile protectionists and the urban English Proletariat
Concept of money	Substance and Mixed Mode	Notion	Convention
Genre	Pamphlet	Query Series	Essay
Impact of money	The Mother of Civil Society	Part of B's "second conceptual revolution"	The philosophy of commercial society

5. *Locke as a monetary philosopher: substance and mixed mode ideas are the foundation of his concept of money and monetary phenomena are crucial to his theory of society.*

John Locke was a philosopher who was definitely committed to a monetary program precipitated by the so-called "Recoinage Crisis" of 1696. He, along with a number of other experts (including his friend, the Master of the Mint, Isaac Newton), was asked by Sir John Somers to present his recommendation to deal with the fact that the English nation's silver coinage had experienced a dramatic decline in silver content during the previous 3 years. Something had to be done with the coinage since its deterioration was affecting the government's power to borrow money in the midst of a war. William Lowndes, the Secretary of the Treasury, called for a recoinage that would make the *de facto* devaluation of the currency in terms of silver content *de jure*.[14]

Locke entered this contested territory with a clear program directed against Lowndes' proposal: the clipped coins in circulation should be returned to the mint and be exchanged (by weight) with the newly minted full value coins ("full weight recoinage"). He argued that Lowndes' devaluation or "inflationist" solution would in effect reward the coin clippers, baggers and counterfeiters and put them in a position of dictating the monetary policy of the most powerful nation on the planet. Such a resolution was an anathema for Locke, but he also recognized that his policy recommendation posed the threat of dramatic deflation in the midst of a war.

He used his philosophical apparatus to defend his recommendation and resolve this dilemma by defining money in such a way that the consequences of not going through with a "full weight recoinage" were evidently much more dangerous than the threats of deflation and economic crisis posed by going through with it. For Locke points out that in exchanges "[silver] is the thing bargain'd for, as well as the measure of the bargain."[15] It operates partially as a desired substance, but when silver is coined, a new ontological element is added, what Locke called a mixed mode: "The Coining of Silver, or making Money of it, is the ascertaining of the quantity by a public mark, the better to fit it for Commerce."[16]

The idea of money then has a two-fold ontological character: a substance idea and what Locke called a "mixed mode" idea.[17] Together they form a very subtle (and ever potentially duplicitous) idea that spans nature and culture and synthesizes physical primary qualities with language-based unions of ephemeral acts, thoughts and events. Locke was clear as to the contrast between the two ontological categories: "Besides the greatest part of mixed modes, being actions which perish in their birth,

[14] For a thorough account of the historical setting of the recoinage crisis see Kelly (1991).

[15] Kelly (1991), 412.

[16] Kelly (1991), 412.

[17] Patrick Hyde Kelly and I in separate works – *Locke on Money* and *Clipped Coins, Abused Words and Civil Government: John Locke's Philosophy of Money* respectively – offered different versions of this two-fold analysis of Locke's conception of money.

are not capable of lasting duration, as substances which are the actors."[18] Is it any wonder then that money has a mysterious character?

For Locke, however, since the study of money involves mixed modes that define real essences, it allows for a form of demonstrable reasoning as strict as that found in mathematics (even though the ideas are less orderly). Indeed, monetary reasoning operates in the disjunctive realm that includes both the moral (*practica*) and the physical (*physica*). That can explain why his monetary writings are rife with moral conclusions – "robbing the Honest Man," "defraud the King," "totally destroy the publick Faith," "Clipping and false Coining hightens the Robbery into Treason."[19] – derived with demonstrative reasoning. For moral reasoning, being about humanly defined ideas, can have deductive consequences. For example, he writes: "'Where there is no property, there is no injustice,' is a proposition as certain as any demonstration in Euclid: for, the idea of property being a right to anything, and the idea to which the name 'injustice' is given being the invasion or violation of that right…"[20] Indeed, the reason why money can be studied deductively is not due to its substantial physical aspect; it is due to its being a mixed mode.

Once one understands the metaphysical complexity of money, one understands the vital importance (and difficulty) of coordinating its substance and mixed mode aspects carefully. Indeed, Locke's defense of his monetary program – full weight recoinage – was based on this kind of reasoning; he argued that the threat of monetary deflation was less problematic than the possibility of undermining of the entire monetary system. Both the criminal clippers and the legal devaluationists (or "inflationists") undermine the mutual trust between citizen and state required for the functioning of a modern monetary system: "Altering the Standard, by raising the money … will weaken, if not totally destroy the publick Faith, when all that had trusted the Publick … shall be defrauded by 20 per Cent."[21] I.e., a monetary crisis (like deflation) is infinitely less dangerous than a crisis of the money form itself.

Locke's program and the impact of his philosophy on the concept of money was definitely powerful, but what of the impact of money on Locke's philosophy? Is there evidence of this affect? I argue that there is and that it is best seen in Locke's *Second Treatise on Government*, where he literally defines both a period of human history as well as a social stratum that is always already in existence with the realm of money. Historically, money drove the transition from the period of subsistence to the emergence of a fully developed system of accumulation. Moreover, the contemporary world of international trade that operates in a pre-civil society manner is based on money (especially its substantial element).

Money, for Locke, is the mediator between nature and culture in history as well as metaphysically. This mediation arises from an unintended consequence to the

---

[18] Locke (1979), III, v, 8.

[19] Kelly (1991), 417, 417, 417, 415, respectively.

[20] Locke (1979), IV, iii, 18.

[21] Quoted in Kelly (1991), 417.

tacit agreement to accept gold and/or silver as the universal equivalent taken by an overwhelming majority: the end of the moral obligation to share the surplus and to provide a "potlatch," if one can. This sets the stage for the development of the accumulation process, the privatization of land and eventually the whole system of civil government. So, for Locke, money is "The Mother of Civil Government."[22]

Thus we see that Locke is, according to my definition, a philosopher of money in that he proposes a monetary program, he uses his philosophical apparatus as a tool for the formation and defense of his proposal and his philosophical work is deeply affected by his money environment.

6. *Berkeley as a monetary theorist: the critique of substance, primary qualities, and objectivity and the role of money in his "second conceptual revolution."*

It should not come as a surprise that John Locke was a philosopher of money. Marx, for example, characterized Locke's role in the recoinage crisis in the following words: "John Locke, who championed the new bourgeoisie in every way … even demonstrated in a separate work that the bourgeois way of thinking is the normal human way of thinking – took up Lowndes' challenge."[23] But it may come as a surprise that Berkeley was also a philosopher of money.

Berkeley certainly had a monetary program that he proposed and actively supported.[24] It called for the creation of a National Bank and a National Mint for Ireland. However, the Bank would issue paper money while the Mint will coin "small change" tokens with no trace of specie content in them. He proposed this plan with the help of members of the "improving" Dublin Society and in defense of the tithes and rents of the Church of Ireland (that alliance indicates Bishop Berkeley's complex socio-political position). Berkeley was also clear as to his opposition: the poor, but cynical Irish "natives" who refused to work and the wealthy, libertine-influenced rural gentry who refused to pay their tithes and the British mercantile class that refused to let Ireland be. The whole point of his monetary program was to excite the Irish natives into productive labor, to force the wealthy gentry to spend their income (derived from rents and profits) "at home," and to liberate the Irish economy from the control of the Parliament in London.

Berkeley deployed his philosophy in the development and defense of the form of paper money he was advocating barely 15 years after the crash of the great experiment with specieless money: the Law System. This was not an easy task and he definitely needed to attack the notion of money that his great predecessor, John Locke, presented. Key to Locke's concept was the necessity for the substantial aspect of money. Berkeley suggested in the voice of the *Querist* that there was no such necessity. On the contrary, Locke's prejudice for a metallic substance element in the idea of money was an invitation to intellectual and political catastrophe for Ireland.

[22] Caffentzis (1989), 70.
[23] Marx (1970), 77.
[24] My account here is based on Caffentzis (2000).

Berkeley criticized the view that material substances had a level of objectivity that notions (the self, its powers and activities) lack. Moreover, he questioned the necessity for a material basis of the monetary system. For the essential function of money is not to represent a valuable thing or substance, but it is to excite productive action. The presence of gold or silver substances in a coin was no "guarantee" as to its objective value and worth *qua* money. For money is fundamentally notional and has no attachment to material substance ideas that are intellectually incoherent anyway. For gold's and silver's primary quality character (that supposedly gives them the virtue of an enhanced objectivity) is open to decisive mental relativity arguments by which Berkeley shows that "primary" qualities are as relative to the mental agent as are "secondary" ones.

This critique of Locke's concept of money and Berkeley's subsequent defense of a specieless currency called on all aspects of his philosophical system. Did money also affect Berkeley's philosophy as well? I believe that it did. For Berkeley's philosophy until the 1730s was divided between ideas and selves. Ideas were dependent on selves, but selves and their faculties and actions were undescribable by ideas.

This dichotomy made for a powerful critical machine when aimed at materialist doctrines, however it left Berkeley unable to describe or plan for a world of selves. Berkeley went through a "second conceptual revolution" in the 1730s through the introduction of the notion (and its cognates) that allowed him to do exactly that. As I wrote in *Exciting the Industry of Mankind*:

> For Locke, money was a complex idea bringing together material substance ideas with mixed-mode notions, whereas for Berkeley money was a mixed-mode notion stripped of any essential dependence on material substances. Its purpose was to stimulate and regulate action, not to measure and store a quantity or specie.[25]

The Querist was a vital part of Berkeley larger intellectual revolution, for in differentiating between his and Locke's view of money, Berkeley was able to apply his thought onto collectivities of selves, describe their momentum, and plan their future trajectories.

The inspiration for this effort, I believe, arose from his experience in the American colonies, especially his two and a half years of life in Rhode Island, the epicenter of a great monetary experiment in defetishizing specie that was taking place.[26] As I wrote: "[Berkeley] saw in Newport a society which had transcended the metallic gravity of coinage to survive and even flourish."[27] In this American context there was also another factor, for the Rhode Island colony was "surrounded by a sea of gift-exchange relations among the native Americans...the colonists' wampum transactions with the native Americans liberated their monetary imagination just as the masterless character of the North American Indian peoples liberated the colonists' political imagination."[28]

---

[25] Caffentzis (2000), 274.

[26] Caffentzis (2000), 80–100.

[27] Caffentzis (2000), 82.

[28] Caffentzis (2000), 418.

Thus George Berkeley was a philosopher of money. He proposed and fought for a monetary program (that, however, failed in the short run, but that eventually triumphed); he deployed his philosophical thought in its defense and development, and finally his philosophy was deeply affected by monetary experiences.

7. *Hume as a monetary theorist: a reaction to the '45; the importance of convention as the basis of monetary exchange; money and monetary phenomena make the creation of history possible (i.e., Hume's recognition that the future will not be determined by the past).*

Berkeley was dealing with the problem of having the finances of the Church of Ireland depending on an Irish Catholic working class that was passive and resistant to work. His conception of money was dependent upon this problematic. Hume published his *Political Discourses* in Scotland 7 years after one of the most momentous events in his country's history: the almost successful Jacobite rebellion against the King called the '45er. The enlistment of thousands of Highlanders (and many Lowlanders) from both the elite and the "common people" of Scotland into the army of the rebellion showed that the 1707 Act of Union between Scotland and Britain had not succeeded in convincing Scots that it marked a path to their economic betterment. Scotland could no longer remain a poor country and Hume, a leader in the growing circle of Scottish Enlightenment thinkers, had to show the path out of poverty. Hume's 1752 *Political Discourses* was his "What is to be done?" text, where he proposes a monetary program (as well as a critique of other monetary strategies). His key idea was rooted in a response to the so-called "rich country-poor country" debates that had their origin in the post-Glorious Revolution era: if a poor country (with low wages and cost of living) was inserted in the midst of world trade flows and was hindered neither by geographical barriers nor tariff walls, it would inevitably be able to successfully compete with the rich countries, due to the low price of the commodities it produces.[29]

The proper monetary policy to promote this "natural" advantage that a poor country has is one that opens it up to the flows of money that participation in world trade invites. The introduction of surplus money due to the positive balance of trade induced by the competitive edge of the poor country's low wages will stimulate the industry of workers and capitalists and thus further drive them on the path of commercial development. Hume was hostile to any program like George Berkeley's that called for withdrawing from world trade into autarky and for substituting local paper money created by a national bank for world money (which at that time was specie-based coinage). Autarky and paper money, Hume argued, would doom a country to an eternal sentence of poverty and misery. In the case of Scotland, it would also inflame the passions that would ignite futile rebellion again and again.

Hume's answer to Scotland's plight required that he promote world money (gold and silver coinage) and criticize the invention of new paper financial instruments, especially those based on credit and debt. This position was surprisingly reactionary

[29] Hont (2008), 243–327.

and might appear a bit paradoxical given his other philosophical commitments. Hume, however, used his philosophy to support his monetary program and criticize his opponents. One of the most salient examples of this effort is to be found in his differential approach to metallic versus paper money.[30] Hume argued that the use of paper money issued by a National Bank is a catastrophe waiting to happen and in a society like Scotland, where the refinement and discipline required to preserve trust in a commercial world were lacking, the wait would not be long. Why this hostility to paper money? Hume, after all, was as clear as Berkeley about the conventionality and "fictionality" of money in general, so why did he accuse paper money in particular to be "counterfeit"?

The answers to these questions lie in Hume's philosophical analysis of fictions and conventions. For Hume distinguished between natural and artificial fictions and between conventions and promises. Natural fictions are those that arise in the construction of concepts like duration through time, continued existence of objects, self, substance, and ideal standards. They operate through a universal mental propensity of confounding "identity with relation." Though strictly they are not arrived at either by deductive or even inductive methods they provide, as Annette Baier characterizes them, "plausible stories we tell ourselves to organize our experience."[31]

Artificial fictions, however, are not universally believed. They are the equivalent of constructed illusions whose success is dependent upon the idiosyncratic gullibility of individuals. They can be found in art, literature, myth and religion and not only in circuses and magic shows.

Hume also distinguishes between conventions and promises. Both are products of human exchanges, but the former is much more basic and reliable than the latter. A convention arises from the interlinked activities of at least two persons whose logic is: "Whatever is advantageous to two or more persons if all perform their part, but what loses all advantage if only one person perform…."[32] Conventions are the ur-relations that form society, promises and contracts come later. Hume includes the collective activities (like rowing), money and language under rubric of convention: "Thus two men pull the oars of a boat by common convention, for common interest, without promise or contract; thus gold and silver are made the measures of exchange; thus speech, and words, and language are fixed by human convention and agreement."[33]

Consequently, we can see some basic philosophical distinctions arising between metallic and paper money. For metallic money "shares with time, ideal standards, substances, endurance of sensed objects, and the self a fundamental 'pathos of identity,' that is, every coin reflexively says of itself to both buyer and seller that it has a certain intrinsic value," while paper money clearly does not have the character of a

---

[30] My discussion of Hume's support of metallic money and his critique of paper-credit, as he called it, can be found in Caffentzis (2008), 146–167.

[31] Baier (1991), 103.

[32] Hume (1957), 123.

[33] Hume (1957), 123.

"natural fiction."[34] Also, coinage has the original character of conventional exchange – I give you this coin, if you give me this apple/I give you this apple, if you give me that coin – while paper money has the quality of a promise that requires a threatening deferral in time. Thus paper money is an artificial fiction based on a promise while metallic money is a natural fiction based on a convention.

This philosophical distinction between metallic and paper money gave Hume the intellectual support to challenge Berkeley's autarky and paper money solution to the rich country – poor country problematic. But was Hume's philosophy affected by the monetary forces of his day? From his earliest works to his final efforts, there is a constant emphasis on the importance of money not as a thing, but as a set of social relations in his philosophy. Hume saw in the increasing intensity of monetary exchange not just a sign of increasing wealth, but also of industry, art and refinement, i.e., civilization. His philosophy was therefore a reflection on money and its multiple shadows. As Carl Wennerlind, a scholar of Hume's work, concludes in a recent study of Hume's theory of money:

> Money also plays a central role in Hume's political philosophy. As one of the three foundational conventions of the modern social form [alongside property and markets], Hume suggests that money is a necessary condition for the establishment of a civilized, prosperous, and liberal society.[35]

Hume, therefore, satisfies the final criterion, clearly making him a philosopher of money.

8. *Conclusion: An apology vindicated?*

The defense of my definition of what it is to be a philosopher of money and its application in the cases of Locke, Berkeley and Hume is now complete. I recognize, however, that there is a remaining, but additional question that could only be asked at the conclusion of this paper: what is the value of such an effort? Clearly, the ultimate answer to this question is not mine to give. But I can make one observation as to why this work has been valuable to me (besides the pleasure of researching and writing it).

I also have found that working on this trilogy has made it possible for me to more adequately understand the great monetary transformation that was taking place (with much struggle and confusion) in the latter part of the seventeenth and throughout the eighteenth centuries in Europe by fits and starts (and that "ends" in August 1971): the development of a specieless world money system. This transformation required not only a change in the concept of money; a change in the self-understanding of money users was also required. This happened neither automatically nor ironically. But philosophers of money played an important role in the creation of this new self-understanding. I have no doubt that when, or if, the Grand Narrative of this process is ever written, the conflicting and conflicted work of Locke, Berkeley and Hume will be central to telling the tale.

[34] Caffentzis (2008), 164.

[35] Wennerlind (2008), 124.

References

Baier, A. *A Progress of the Sentiments.* Cambridge: Cambridge University Press, 1991.

Blaug, M. *Economic Theory in Retrospect.* Homewood, Il: Richard D. Irwin, 1968.

Blaug, M. *et al. The Quantity Theory of Money: From Locke to Keynes and Friedman.* Hants, England: Edward Elgar, 1995.

Caffentzis, C.G. *Clipped Coins, Abused Words and Civil Government: John Locke's Philosophy of Money.* Brooklyn: Autonomedia, 1989.

Caffentzis, C.G. *Exciting the Industry of Mankind: George Berkeley's Philosophy of Money.* Dordrecht: Kluwer, 2000.

Caffentzis, C.G. Hume, Money, and Civilization; Or, Why Was Hume a Metallist? *Hume Studies,* 27(2) (2001): 301–335.

Caffentzis, C.G. Civilizing the Highlands: Hume, Money, and the Annexing Act. *Historical Reflections,* 31(1) (2005): 169–194.

Caffentzis, C.G. Fiction or Counterfeit? David Hume's Interpretations of Paper and Metallic Money in *David Hume's Political Economy,* edited by Carl Wennerlind and Margaret Schabas, 146–167. London: Routledge, 2008.

Derrida, J. *Given Time: I. Counterfeit Money.* Chicago: University of Chicago Press, 1992.

Hont, I. The "Rich Country-Poor Country" Debate Revisited: The Irish Origins and French Reception of the Hume Paradox in *David Hume's Political Economy,* edited by Carl Wennerlind and Margaret Schabas, 243–327. London: Routledge, 2008.

Hume, D. *Inquiry Concerning the Principles of Morals.* Indianapolis: Bobbs-Merrill, 1957.

Johnston, J. *Bishop Berkeley's Querist in Historical Perspective.* Dundalk: Dundalgan, 1970.

Kelly, P.H. *Locke on Money,* Vols. I and II. Oxford: Clarendon, 1991.

Locke, J. *An Essay Concerning Human Understanding,* edited by Peter H. Nidditch. Oxford: Oxford University Press, 1979. Reference is to Book, Chapter, and Section.

Marx, K. *Contribution to a Critique of Political Economy.* New York: International Publishers, 1970.

Shell, M. *Money, Language, and Thought: Literary and Philosophical Economies from the Medieval to the Modern Era.* Berkeley: University of California Press, 1982.

Simmel, G. *The Philosophy of Money.* London: Routledge, 2000.

Spiegel, HW. *The Growth of Economic Thought.* Durham, NC: Duke University Press, 1983.

Wennerlind, C. An Artificial Virtue and the Oil of Commerce, *David Hume's Political Economy,* edited by Carl Wennerlind and Margaret Schabas, 105–126. London: Routledge, 2008.

Chapter 6
Berkeley and Chemistry in the *Siris*

The Rebuilding of a Non-existent Theory

Luc Peterschmitt

In this paper, I would like to show how it is possible to understand and comment on Berkeley's *Siris*. This book is not that difficult nor that obscure. *Siris* is unusual: Berkeley seems to have or to invent a new philosophical style. However, firstly, it is still philosophy; and, secondly, it is necessary to stress that, unlike his first works, *Siris* was read everywhere in Europe.[1] And its success is not based only on the success of tar-water – for example, in 1752, in the French translation of an English translation of a French book, Nicaise Le Febvre's *Traicté de Chymie*, the translator added a translation of the first letter to Prior about tar-water, where Berkeley expounds his most speculative conceptions of the universe and the soul of the world.[2] However, the cause of the success of *Siris* in the middle of the eighteenth century is nowadays the cause of its difficulty: it depends on a context which has been quite forgotten. To summarize these considerations, this difficulty is not a difficulty of the text in itself: it is an effect of our ignorance. This does not imply

---

[1] See S. Charles, "La *Siris* au siècle des Lumières: panacée ou imposture?", *Hermathena*, 2000, pp. 55–69.

[2] See N. Le Fébvre, *Cours de chymie, pour servir d'introduction à cette science*, par Nicolas Lefevre (sic), Professeur Royal de chimie et Membre de la Société Royale de Londres, cinquième édition, revue, corrigée et augmentée d'un grand nombre d'opérations et enrichie de figures par M. du Monstier, Apothicaire de la Marine et des Vaisseaux du Roi, Membre de la Société Royale de Londres et de celle de Berlin, 5 vol., Paris, 1752, vol. 2, pp. 437–455. N. le Febvre was a Protestant chemist, emigrated in the United Kingdom in the 1650s, who became a fellow of the early Royal Society.

L. Peterschmitt (✉)
118 avenue de Boufflers, 59130 Lambersart, France
e-mail: luc.peterschmitt@wanadoo.fr

S. Parigi (ed.), *George Berkeley: Religion and Science in the Age of Enlightenment*,
International Archives of the History of Ideas/Archives internationales d'histoire des idées 201,
DOI 10.1007/978-90-481-9243-4_6,

that *Siris* is not a serious book, nor that it is unimportant – especially as far as philosophy of science is concerned.[3]

It is impossible to prove all of this in this paper. I will only comment on Section 202 of *Siris* in order to illustrate the general position I have just expressed. It is easy to explain this choice:

1. Berkeley's argument in this section is quite easy to understand, and the role of this section in *Siris* is not difficult to establish.
2. As Berkeley mentions his sources, which he does not always do, it is not difficult to find them: the context is almost immediately clear.

First, I will closely examine Berkeley's argument. Then I will explain its role in *Siris.* In the last section of my paper, I will show how Berkeley works to elaborate his argument, and what kind of difficulties it presents.

6.1 Section 202 of *Siris*

Here is Section 202:

> But thus much seems plain, that whatever is ascribed to acid may be also ascribed to fire or æther. The particles of æther fly asunder with the greatest force: therefore, agreeably to Sir Isaac Newton's doctrine, when united they must attract each other with the greatest force. Therefore, they constitute the acid. For whatsoever strongly attracts and is attracted, may be called an acid, as Sir Isaac Newton informs us in his tract *De Acido.* Hence it should seem that the sulphur of Homberg and the acid of Sir Isaac are at bottom one and the same thing, to wit, pure fire or æther.[4]

Berkeley's argument runs as follows:

1. Whatever is ascribed to acid may be also ascribed to fire or æther.
2. (Therefore) acid is fire or æther.
3. The particles of æther fly asunder with the greatest force.

[3] Berkeley proposes quite important insights about the philosophy of chemistry, but I cannot develop them here; however, my reading of *Siris* is completely opposed to the way in which I. Tipton or A. A. Luce, for example, read it: see I. Tipton, "Two questions on Bishop Berkeley's Panacea", *Journal of the History of Ideas,* 30, 1969, pp. 203–224; A. A. Luce, "Berkeley's Search for Truth", *Hermathena,* 82, 1953: "I certainly wish he had not spent all that time on *Siris*; the part on tar-water justified itself; the rest is scaffolding, and never should have seen the light of the day as such" (p. 22). About Berkeley's philosophy of chemistry, see my "Berkeley et les principes de la chimie: des lois pour la chimie?", *in* B. Joly (ed.), *La chimie dans l'œuvre des philosophes,* Oxford: Kluwers "Cahiers de logique et d'épistémologie" (forthcoming). According to Berkeley, chemistry does not provide causal explanations, but rests on particular laws describing the motions of sub-microscopic particles. Chemistry is a science of particular phenomena: it has to give particular explanations.

[4] *Siris,* § 202, in *The Works of George Berkeley,* ed. by Luce and Jessop, 9 vols, London and Edinburgh: Nelson, 1948–1957, vol. 5, p. 100.

4. (Therefore) when united they must attract each other with the greatest force.
5. Whatsoever strongly attracts and is attracted, may be called an acid.
6. (Therefore) the particles of æther constitute the acid.
7. The acid of Sir Isaac Newton is pure fire or æther.
8. The sulphur of Homberg is pure fire or æther.
9. (Therefore) it should seem that the sulphur of Homberg and the acid of Newton are at bottom one and the same thing

It seems that the whole argument is centered on conclusion 2, which is drawn from proposition 1. Such a conclusion is valid if one adopts the chemical definition of the nature of a body: the chemical nature is determined by the properties of the body. This is the way one reflects when testing if a metal is gold, as Boyle puts it. More generally, chemists identify bodies in this way. For example, one reason Homberg thinks that vitriolic acid is the same as the acid made from sulphur is the fact that "everything which is done by the spirit of vitriol can be also done by spirit of sulphur, and reciprocally".[5]

Then Berkeley has to justify proposition 1. He does it by working with a fundamental property of the particles of acid and æther: their attraction. This does not mean that Berkeley admits in *Siris* the existence of corporeal forces. As in *De Motu*, and even in the *Principles*, he interprets attraction as a certain motion of particles, which tend to approach one another.[6] Such an interpretation is quite close to Newton's own position – at least to some of his declarations. Newton declares in his *Principia Mathematica*:

> I use the word 'attraction' here in a general sense for any endeavour whatever of bodies to approach one another, whether that endeavour occurs as a result of the action of bodies either drawn toward one another or acting on one another by means of spirits emitted or whether it arises from the action of æther or of air or of any medium whatsoever – whether corporeal or incorporeal – in any way impelling toward one another the bodies floating therein.[7]

Last, this presentation of Berkeley's argument shows how it relies on authorities: propositions 4 and 6 are concluded respectively from propositions 3 and 5, "agreeably to Newton's doctrine".

---

[5] Homberg, "Essai de l'analyse du soufre commun", *Mémoires de l'Académie Royale des sciences*, 1703, p. 39: "Tout se qui se fait par l'esprit de vitriol se peut faire de meme par l'esprit de soufre, et vice-versa" (in this paper, all translations from French and Latin into English are mine).

[6] See for example, *Principles of Human Knowledge*, § 103, *op. cit.* (vol. 2, p. 86): "The great mechanical principle now in vogue is attraction. That a stone falls to the earth, or the sea swells towards the moon, may to some appear sufficiently explained thereby. But, how are we enlightened by being told this is done by attraction? Is it that that word signifies the manner of the tendency, and that it is by the mutual drawing of bodies, instead of their being impelled or protruded towards each other?".

[7] *Mathematical Principles of Natural Philosophy*, section XI, scholium, eng. transl. by I. B. Cohen and A. Whitman, Berkeley and Los Angeles: University of California Press, 1999, p. 588; see also the definition of the motive quantity of centripetal force: "this concept is purely mathematical, for I am not now considering the physical causes and the sites of forces" (Def. 8, p. 407). Many parallel texts can be found in Newton's works.

6.2 Why Does Berkeley Try To Reconcile Newton and Homberg?

Before making precise comments on this argument, it is necessary to ask: why does Berkeley develop such an argument? What is his purpose in this Section 202? To state it in another way: why prove that Newton's acid and Homberg's Sulphur are the same thing, when this is obviously false? Some general remarks will be sufficient to understand Berkeley's point. Berkeley's evident purpose in *Siris* is to give a natural theology, starting from tar-water and tracing back the chain of being to God. It is easy to explain this way of arguing: Berkeley wants to prove the benevolence of God, who has given to men a panacea that is easy to produce. However, this is only a part of the argument. Indeed, Berkeley tries to explain the properties or virtues of tar-water. In addition, this is a second way of arguing: he seeks for the principle that would account for his panacea. This research leads Berkeley to discover or to show the divine order of the world: God uses one instrument, the Fire – I mean the principle, and not the vulgar or common fire. Last, men can use this instrument for their own purposes – when they use the common fire or a burning-glass. Much more could be said about Berkeley's theology in *Siris*, but that is enough for my purpose. The principle of Berkeley's natural theology is this: he relies on chemistry, rather than on mechanics and astronomy because he found that chemistry gives a more convincing basis for such a theology (in a word, chemistry deals with familiar objects, and consequently shows the action of God where it really interests us, on Earth[8]). Divine action is shown through the role of its instrument: Fire.

However, this is the very point where things become difficult. In the beginning of the eighteenth century, there was not one unified chemical theory. There was no Newton to give chemistry its "synthesis".[9] Many different chemical theories existed. This situation gave rise to debates and polemics. In addition, one can doubt that chemistry is a good basis or foundation for a natural theology (it is not possible to prove divine Providence convincingly from a basis on which nobody really agrees: all the arguments are in that case disputable). Berkeley had two possibilities: (1) to choose one chemical theory, but in this case the choice would have been disputable, and the theology would not have been a solid one; or (2) to make this synthesis. This latter is Berkeley's way in *Siris*. In general, Berkeley relies on the use of authorities, trying to see where they agree. The agreement between recognised authorities is taken as a sign of truth. This way of reasoning leaves Berkeley the possibility of being critical, as I will show: he is not committed to being Newtonian or anything else. Berkeley refers mainly to three chemists who were

---

[8] For a complete account, see my "Philosophie chimique et théologie naturelle dans la *Siris* de Berkeley", *Revue du 18ème siècle* (*Revue du Dix huitième siècle*, 42, 2010, pp. 417–432).

[9] A. Koyré, "Sens et portée de la synthèse newtonienne", *in Etudes newtoniennes*, Paris: Gallimard, 1968.

leading natural philosophers: Newton, Boerhaave, who was a chemist and physician in Leiden, and Homberg, a leading chemist of the Parisian Royal Academy until his death in 1715.[10]

Nevertheless, to reconcile Newton and Homberg about the first principle of nature, given the fact that Homberg and Boerhaave seem to agree on the status of Fire, amounts to showing that according to the best chemistry, Fire is the first principle and may be considered as the main instrument of God.[11] It seems that Berkeley then has a solid basis for his natural theology – that is to say, a basis on which the best chemists do agree.

According to Newton and Homberg respectively, acid and Fire are at least principles of the first importance, if not the first principle(s). As to Homberg, it is evident, and Berkeley is quite aware of that fact:

> Salts are vulgarly reckoned the most active of chemical principles. But Homberg derives all their activity from the sulphurs joined with them.[12]

That corresponds to Homberg's position:

> These principles are of three different natures; that is to say: one active principle, a passive one and three middle ones. The active principle is Sulphur, the passive is Earth, and the middle ones are Salt, Water and Mercury. I call Sulphur the active principle because it acts alone and make the other acts.[13]

Moreover, it is necessary to add that the only time Homberg speaks of God in his work is in the conclusion of his paper on Sulphur. He notes that the matter of light (or Fire or Sulphur) can change the substance of bodies and then produce an infinite diversity of bodies. This production of bodies proves, according to Homberg, that with the same first principles, it would be possible to produce other worlds than ours:

> Even if there could be several worlds as ours, they could be all furnished with diverse objects, without changing the matter not the manner in which these objects would be

---

[10] About Boerhaave, see G.A. Lindeboom, *Herman Boerhaave, The Man and his Work*, London: Methuen & co, 1968; about Homberg, and its importance, see Saint-Simon's judgement: "The duke of Orléans lost meanwhile Homberg, one of the greatest chemists in Europe", *Mémoires* (year 1715), Paris: Gallimard, 8 vols, 1983–1988, vol. 7 p. 742. Both Boerhaave and Homberg are very important figures in the history of chemistry. I should add some other natural philosophers, of less importance in *Siris*, as Nehemiah Grew, Stephen Hales, and Bernard Nieuwentyt.

[11] This notion of an element which is also an instrument fits well to Boerhaave's conception of elements – see R. Love, "Herman Boerhaave and the Element-Instrument Concept of Fire", *Annals of Science*, 1974, pp. 547–559, and my "Philosophie chimique et théologie naturelle dans la *Siris* de Berkeley".

[12] *Siris*, § 136, *op. cit.*, vol. 5, p. 77.

[13] Homberg, "Essais de chimie. Article premier. Des principes de la Chimie en général", *Mémoires de l'Académie Royale des Sciences*, 1702, p. 34: "ces principes sont de trois différentes natures; savoir un principe actif, un passif, et trois moyens. Le principe actif est le soufre, le passif est la terre, et les principes moyens sont le sel, l'eau et le mercure. Nous appelons le soufre principe actif parce qu'il agit seul et fait agir les autres."

> composed; this proves the infinite richness and power of the Being who produced the universe.[14]

As to Newton, it is necessary to examine the role that he attributes to acids. According to Berkeley:

> It should seem rather that the vehement attraction which Sir Isaac Newton attributes to all acids, whereby he supposeth them to rush towards, penetrate, shake, and divide the most solid bodies, and to ferment the liquid of vegetables, could better account for this phenomenon. It is in this attraction that Sir Isaac placeth all their activity.[15]

This is quite correct. To Newton, the attraction with which the particles are "endowed" is the cause of fermentation:

> By this attractive force they [the particles of acids] get about the particles of bodies, whether they be of a metallick or stony nature, and adhere to them most closely on all sides; so that they can scarce be separated frome them by distillation or sublimation. When they are attracted and gather'd together about the particles of bodies, they raise, disjoin and shake them one from another; that is they dissolve those bodies. By their attractive force also, by which they rush towards the particles of bodies, they move the fluids and excite heat; and they shake asunder some particles, so much as to turn them into air, and generate bubbles: and this is the reason of dissolution and all violent fermentations.[16]

Once again, the acids, as the cause of fermentation, play a crucial role in the world:

> Seeing therefore the variety of Motion which we find in the World is always decreasing, there is a necessity of conserving and recruiting it by active Principles, such as are the cause of Gravity, by which Planets and Comets keep their Motions in their Orbs, and Bodies acquire great Motion in falling; and the cause of Fermentation, by which the Heart and Blood of Animals are kept in perpetual Motion and Heat; the inward Parts of the Earth are constantly warm'd, and in some places grow very hot; Bodies burn and shine, Mountains take fire, the Caverns of the Earth are blown up, and the Sun continues violently hot and lucid, and warms all things by his Light.[17]

The role of acids is then cosmological: it offsets the effects of the mechanical laws of motion, which are potentially destructive. If it could be shown that Homberg's Fire and Newton's acid are the same thing, then this thing would be a very interesting first principle for a natural theology.

---

[14] Homberg, "Suite de l'article trois des Essays de chimie", *Mémoires de l'Académie Royale des Sciences*, 1706, p. 272: "S'il pouvait y avoir plusieurs mondes comme le nôtre, ils pourraient tous être différemment garnis d'objets sans changer la matière, ni la manière dont ces objets seraient composés; ce qui marque une richesse et une puissance infinie de l'Être qui a produit l'univers".

[15] *Siris*, § 133, *op. cit.*, vol. 5, p. 76; see also *Siris* §§ 240, 250.

[16] Newton, *De Natura acidorum*, *in* John Harris, *Lexicon Technicum*, reprint *in Isaac Newton's Papers and Letters on Natural Philosophy and related Documents*, ed. by I.B. Cohen and R.E. Schofield, Cambridge: Cambridge University Press, 1958; eng. transl. rev. by Newton, *Some Thoughts about the Nature of Acids*, *ibid.*, p. 257 (hereafter *De Natura Acidorum*).

[17] Newton, *Opticks: Or, A Treatise of the Reflections, Refractions, Inflexions and Colours of Light. The Second Edition, with Additions*, London: 1717, Question 31, p. 375 (hereafter *Opticks*).

6.3 The Rebuilding of a Non-existent Theory

Berkeley's method in *Siris* leads him to build a kind of a historical monster: a chemical theory that is supposed to be *the* chemical theory, but which does not exist anywhere except in *Siris*. One may say that what Berkeley takes to be the common chemical theory is not even false; it is instead Berkeley's own interpretation of the various chemical theories, and nothing else. Indeed, it is not possible to reconcile Newton and Homberg. They just do not agree.

I will now show that Berkeley's argument is not correct from a historical point of view. To this purpose, it is enough to compare what he makes Newton say and what Newton says or does not say. I will closely examine propositions 4, 5 and 7.

Let us begin with proposition 5: the fundamental property of acids is attraction. This is a quotation from Newton, who declares:

> Whatever doth strongly attract and is strongly attracted, may be call'd an acid.[18]

However, Berkeley uses this "true" premise is a very curious way. To show that æther is acid, or better, constitutes acids, he has to show that the particles of acids attract one another (proposition 4). The problem is that, before this section, Berkeley insists on the elasticity of ether, that is to say that the particles of ether repel one another:

> Fire seems the most elastic and expansive of all bodies. It communicates this quality to moist vapours and dry exhalations, when it heats and agitates their parts, cohering closely with them, overcoming their former mutual attraction, and causing them, instead thereof, reciprocally to repel each other and fly asunder, with a force proportionable to that wherewith they had cohered.[19]

From this elasticity, Berkeley claims to deduce that the particles of ether also attract one another. Berkeley claims that this deduction follows "Newton's doctrine". Indeed Newton affirms:

> And these last are the Bodies which Chymists call fix'd, and being rarified by Fementation, become true permanent Air: those Particles receding from one another with the greatest Force, and being most difficultly brought together, which upon Contact cohere most strongly.[20]

In Section 149, Berkeley correctly quotes Newton; in Section 149, Berkeley needs the reciprocal proposition. However, this converse is not Newtonian. Newton never wrote anything to the effect that what recedes with the greatest force also attracts with the greatest force. In addition, it is not even certain that this reciprocal proposition is compatible with Newton's doctrine. Newton shows that if a coherent body is destroyed, then it is almost impossible to make its parts cohere or unite again. If what is the most coherent is, after destruction, the most repellent, it is not certain that this process can be reversed.

---

[18] *ibid.*, p. 257.

[19] *Siris*, § 149, *op. cit.*, vol. 5, p. 81.

[20] Newton, *Opticks*, Question 31, p. 372.

However, for the sake of argument, let us admit this premise with Berkeley. It is not certain that from 4 and 5, the conclusion follows that æther constitutes acid. Newton says, "Whatever doth strongly attract and is strongly attracted may be call'd *an* acid" (my emphasis). The right conclusion is then that the particles of æther constitute an acid, and not that they constitute the acid. Berkeley means thereby that these particles are the cause of the acidity of the particular acids (they are the bearer of the property "to be acid": that is to say that they are the agent of dissolution). To Berkeley, acids are composed of bodies. This is an extrapolation from Newton's position: according to Newton, there is an "acid part of the spirit of vitriol".[21] But he does not say more than that: does it mean that an acid is composed of particles which are only acid and other particles which determine the kind of acid (for example particles of acid and particles of sulphur joined together make the spirit of vitriol)? Nothing in Newton's published doctrine indicates that it is possible to speak of acid in general as something that would be homogeneous. Newton just indicates that the "particles of acids"[22] are of a middle size, between that of particles of earth and of water, and that they are endowed with a strong attraction. However, to speak of particles of acids in the plural may indicate that there is not only one sort of such particles, even if they share two properties – their size and their power of attraction. It is not certain that Berkeley's position is Newtonian.[23]

From a strictly logical point of view, Berkeley's argument is not really convincing, and leads him far away from a strict Newtonianism, in spite of Berkeley's claim. However, this reading of Berkeley's argument is not the only one. It helps to address the problem, which is to understand why Berkeley admits such discrepancies with Newton's text. My previous remarks do not aim to show that Berkeley is not serious. To conclude this would amount to reading Berkeley's *Siris* as if it were history of science – in this case, surely, it is a bad book, not worth reading. However, Berkeley is not an historian. Moreover, he need not be a faithful Newtonian. He uses Newton's doctrine for his own purpose. The question is now, why did Berkeley believe that such a use was legitimate? Once again, to understand Berkeley's position, it is necessary to examine the intellectual context of *Siris*. It explains Berkeley's motives, that is to say why he believed that it was possible to read the chemists as he read them.

What is to be explained is the identification between Fire and Acid. For this explanation, it is necessary to comment on the function of a dissolvent, as Berkeley does in Section 191:

> As water acts upon salt, or aqua fortis upon iron, so fire dissolves all other bodies. Fire, air, and water are all three menstruums: but the two last seem to derive all their force and

---

[21] *ibid.* p. 353.

[22] *De Natura Acidorum*, *op. cit.*, p. 257.

[23] Maybe Berkeley interprets Newton according to a quite common view, which is also Homberg's position. To Homberg, an acid is composed of pure salt, sulphur and some other matters. Moreover, pure acid is nothing else than pure salt. However if this is Berkeley's reading, it seems that, according to Berkeley, Homberg's acid or salt is the same as his Fire, which would be a complete nonsense. Berkeley proves too much; his rebuilding is, from a historical point of view, an impossible task, once again.

activity from the first (Sect. 149). And indeed there seems to be, originally or ultimately, but one menstruum in nature, to which all other menstruums may be reduced.[24]

According to Berkeley, fire is the cause of dissolution – another way of saying that it is the cause of acidity (which is sometimes understood as the capacity to dissolve other bodies). Even if it is not as evident as in Section 202, Berkeley relies here on Boerhaave, whom he almost quotes:

- As to Fire, Boerhaave declares:

> It is known that it is almost a universal dissolvent, since it liquefies the most part of bodies.[25]

However, we will see that this quotation is to be qualified: when Boerhaave says that fire is "almost" a universal dissolvent, this not only means that it does not dissolve all bodies, but also that it is not really a dissolvent. However, Berkeley can rely on a quotation.

- As to Air, once again, Berkeley makes Boerhaave say a little bit more than what he really says:

> Indeed, in the Air almost all kind of dissolved bodies are moving and are transported, and it is hardly possible that in applying successively such a diversity of particles, some particles would not act as the dissolvent of the body to which they are applied; this body can be rightly said to have been dissolved by a Universal Dissolvent.[26]

This means that Air is not itself a dissolvent, but it contains particles that may act as solvents.

- As to Water, things are clearer, but need a short explanation. Water dissolves bodies; but its strength is a direct function of its heat. Certainly, Boerhaave declares that Water is a dissolvent. That means that it acts as a dissolvent. However, its action is proportional to the quantity of Fire that it contains. To Boerhaave, indeed, Fire is a concurring cause of dissolution, and is even a necessary condition of dissolution. As M. Goupil puts it, "the first of these causes [which concurs to dissolution] is Fire, which melts the bodies, helps to perform many reactions, and increases the restlessness of the particles. The author even adds that it seems to him that the presence of Fire is necessary for the reaction: he relies on the fact that bodies are inert when it is very cold".[27] According to

[24] *Siris*, § 191, *op. cit.*, vol. 5, p. 96.

[25] Boerhaave, *Elementa Chemiae*, Leiden: Joannis Rudolphi, 2 vols., 1732 (hereafter *Elementa*), vol. II, p. 697: "cognoscitur fere solvens universale, quatenus pleraque liquefacit corpora".

[26] *Ibid.* vol. I, p. 540: "Quum [Aer] enim in se gerat, deferatque, fere omnia genera corporum dissolute, fieri vix potuit, quin successive tantam varietam applicando aliquando applicentur quaedam particulae quae idoneae sint illud corpus, instar menstrui, dissolvere; hocque respectu dici poterit fungi vice Menstrui Universalis".

[27] M. Goupil, *Du flou au clair? Histoire de l'affinité chimique de Cardan à Prigogine*, Paris: Editions des travaux historiques et scientifiques, 1991, p. 101.

> Boerhaave, heating Water amounts to making it "stock" particles of Fire. If it does not contain any (particles of) Fire, it is ice, and ice does not dissolve anything. On the contrary, boiling water is Water "saturated" with fire: its dissolving strength is at its maximum. As the nature of Water remains always the same, what is really acting is Fire contained in Water.[28]

Boerhaave's doctrine is perhaps one of the reasons Berkeley had to identify Newton's acid with Fire: this doctrine gives reason to think that Fire is the principle of dissolutions – which must concern acids too.[29] Certainly, it is not a proof; but it certainly is a motive. Moreover, following Boerhaave on this point implies criticizing Newton on another point. That is what explains the identification of æther with Fire that Berkeley makes. I will not comment on this identification in detail. It is just necessary to note that Berkeley refuses Newton's conception of fire and heat as something like a state of matter:

> Sir Isaac Newton in his *Optics* asks, 'Is not fire a body heated so hot as to emit light copiously? For what else', adds he, 'is a red-hot iron than fire?' Now, it should seem that to define fire by heat would be to explain a thing by itself. A body heated so hot as to emit light is an ignited body, that is, hath fire in it, is penetrated and agitated by fire, but is not itself fire. And although it should in the third foregoing acceptation, or vulgar sense, pass for fire, yet it is not the pure elementary fire (Sect. 190) in the second or philosophic sense, such as was understood by the sages of antiquity, and such as is collected in the focus of a burning-glass; much less is it the *vis*, force, or power of burning, destroying, calcining, melting, vitrifying, and raising the perceptions of light and heat.[30]

According to Berkeley, Newton is mistaken when he speaks of Fire: he confounds the vulgar fire and the "pure elementary fire". Therefore, it is possible to identify æther and the pure Fire. It does not directly contradict Newton, since he does not say anything about pure Fire. What interests me in this story is the fact that the way in which Berkeley reads Newton is not completely arbitrary. He gives to Newton's doctrine a Boerhaavian development, so to speak.

However, such a way of reading a text amounts to seeing in it what may be in other texts. I have given only a part of Berkeley's motives. One could ask: but why did Berkeley choose at *this* point to read Newton in a Boerhaavian way, and not to read Boerhaave in a Newtonian manner? In fact, it seems that Berkeley's way of reading was not that unprecedented in the middle of the eighteenth century. Here are some examples of authors trying to elaborate something that is quite analogous to Berkeley's *Siris*. The first two are significant, because they are independent attempts to reconcile on the one side Newton and Homberg, and on the other side Newton and Boerhaave:

---

[28] *Elementa*, vol. II p. 720 sq. – see especially pp. 720–721.

[29] Berkeley may also recall the presentation given by Fontenelle of Homberg's memoir devoted to experiments performed with the help of a new burning glass. Fontenelle writes: "fire has been its universal dissolvent, or almost always the soul of its other dissolvents" (*Histoire de l'Académie Royale des Sciences*, "Sur des expériences faites au miroir ardent convexe", 1702, p. 34: "Le feu a été son dissolvent universel, ou presque toujours l'âme de ses autres dissolvants").

[30] *Siris*, § 221, *op. cit.*, vol. 5, p. 106–107.

1. The author of a very curious manuscript pretends that Newton's æther and Homberg's Fire or Sulphur are one and the same thing, because neither of them is subject to gravity.[31]
2. According to R. Knoeff, "Fahrenheit directly links the universally distributed fire to an ethereal matter".[32] Fahrenheit does not explicitly say that æther and fire are the same thing; but his interpretation leaves this possibility open.

Berkeley did not read these works. However, it does not really matter, since I am not taking them as his sources; they are merely signs that Berkeley's way of reading was quite common in the middle of the eighteenth century.

3. A third and last important text is Hales' *Vegetable Staticks*. According to him, the force of an acid is to be linked with the elastic air that is contained in this acid:

> Have we not reason also hence to conclude, that the energy of acid spirits may in some measure be owing to the strongly attracting air particles in them, which active principles may give an impetus to the acid *spiculae*?[33]

However, Berkeley thinks that what Hales calls "elastic Air" is nothing else than the pure elementary Fire. For my purpose, it will be enough to indicate the main lines of Berkeley's argument. Following Hales, Berkeley distinguishes two parts in the air: an elastic part, which is truly active, and a second part, composed of diverse particles, which may alter the elasticity of air:

> The specific quality of air is taken to be permanent elasticity. Mr. Boyle is expressly of this opinion. And yet whether there be any such thing as permanently elastic air may be doubted, there being many things which seem to rob the air of this quality, or at least lessen and suspend its exertion. The salts and sulphurs, for instance, that float in the air abate much of its elasticity by their attraction.[34]

In this section, Berkeley obviously recalls Hales' analysis:

> Tho' the force of its elasticity is so great as to be able to bear a prodigious pressure, without losing that elasticity, yet we have from the foregoing Experiments evident proof, that its elasticity is easily, and in great abundance destroyed; and is thereby reduced to a fixt state, by the strong attraction of the acid sulphureous particles, which arise either from fire or fermentation: And thereby elasticity is not an essential immutable property

[31] *An Examination of the Newtonian Argument for the Emptiness of Space and of the Resistance of subtile Fluids*, London: T. Cooper, 1740, quot. by R. E. Schofield, *Mechanism and Materialism. British Natural Philosophy in An Age of Reason*, Princeton: Princeton University Press, 1970, p. 107.

[32] R. Knoeff, *Herman Boerhaave (1668–1738), Calvinist Chemist and Physician*, Amsterdam: Koninklijke Nederlandse Akademie van Wetenschappen, 2002, p. 137.

[33] S. Hales, *Vegetable Staticks, Or, An Account of some Statical Experiments on the Sap in Vegetables: Being n Essay towards a Natural history of Vegetation. Also a Specimen of An Attempt to Analyse the Air by a Great Variety of Chymico-Statical Experiments*, London: W. and J. Innys, and T. Woodward, 1732, p. 296. This text is also important because Hales was considered as a great Newtonian – about his alleged Newtonianism, see Desaguliers' review of his book in the *Philosophical Transactions*, vol. 34, 1728, pp. 264–291, and vol. 35, 1729, pp. 322–331.

[34] *Siris*, § 146, *op. cit*, vol. 5, p. 80.

> of air particles; but they are, we see, easily changed from an elastick to a fixt state, by the strong attraction of the acid, sulphureous and saline particles which abound in the air. Whence it is reasonable to conclude, that our atmosphere is a *Chaos*, consisting not only of elastick, but also of unelastick air particles, which in great plenty float in it, as well as the sulphureous, saline, watery and earthy particles, which are no ways capable of being thrown off into a permanently elastick state, like those particles which constitute true permanent air.[35]

However, Hales considers that this elasticity is the fundamental property that accounts for the main properties of Air. In particular, this elasticity is a condition for respiration. Thus it is easy to understand why an air that contains too many particles of sulphur or salt is unbreathable: according to Hales, such an atmosphere is not elastic enough. To him, there is no need to suppose the existence of anything other than a pure elastic air contained in the atmosphere.

Obviously, Berkeley cannot accept such an argument. Indeed, he clearly refutes it:

> That there is some latent vivifying spirit dispersed throughout the air common experience sheweth; inasmuch as it is necessary both to vegetables and animals (Sects. 138, 139), whether terrestrial or aquatic, neither beasts, insects, birds, nor fishes being able to subsist without air. Nor doth all air suffice, there being some quality or ingredient of which when air is deprived it becometh unfit to maintain either life or flame. And this even though the air should retain its elasticity; which, by the bye, is an argument that air doth not act only as an antagonist to the intercostal muscles. It hath both that and many other uses. It gives and preserves a proper tone to the vessels: this elastic fluid promotes all secretions: its oscillations keep every part in motion: it pervades and actuates the whole animal system, producing great variety of effects, and even opposite in different parts, cooling at the same time and heating, distending and contracting, coagulating and resolving, giving and taking, sustaining life and impairing it, pressing without and expanding within, abrading some parts, at the same time insinuating and supplying others, producing various vibrations in the fibres and ferments in the fluids: all which must needs ensue from such a subtle, active, heterogeneous and elastic fluid.[36]

At best, the elasticity of air is a necessary condition for life and respiration. Nevertheless, according to Berkeley it is certainly not a sufficient condition. He attributes to this part of air many more properties than those explained by elasticity alone.[37] However, if elasticity is a necessary condition for life and respiration, then it is possible to retain all that Hales says about air. It is not wrong; it is just incomplete. All that Hales attributes to true and permanent elastic air should be attributed to Fire, which also accounts for what elasticity alone cannot explain. Berkeley can interpret the text I have quoted above in his own way, replacing "elastic air" by "Fire". Once again, such an interpretation is not in accordance with Hales' intentions. However, as these arguments are independent from the argument concerning Newton and Homberg, they give a context in which it is possible to reconcile them.

---

[35] Hales, *Vegetable Staticks*, p. 315.

[36] *Siris*, § 143, *op. cit*, vol. 5, p. 79.

[37] If elasticity could explain every property of air, then it would have to explain opposite effects, which is impossible; Berkeley's solution is to consider that what he calls Fire is in fact composed, just as white light is composed of colours: it can explain even opposite effects.

This context, quickly sketched, shows that Berkeley could believe that his reading of Newton and Homberg was correct. It was a common way of reading – it is not an historical reading of texts, but rather something more philosophical, or at least a creative reading.

To conclude, I wanted to highlight the following points:

- That part of *Siris*, which is devoted to natural sciences and especially to chemistry, is systematic, even if this does not immediately appear. In a word, Berkeley does not follow a logical order, but he follows the order of the elements. He expounds a theory, and he does so in a rational way.[38] Of course, his style here is very different from the style of his first books. But if one works on the scientific context of *Siris*, it is not an obscure book, even if it can be disconcerting for a modern reader.
- Berkeley tries to build a unified chemical theory as a basis for a natural theology. However, Berkeley is not a chemist: I mean thereby that he does not perform experiments, and does not give anything new in chemistry. He intends to synthesize the best chemical theories from the beginning of the eighteenth century – or let's say the theories of the best chemists. Of course this synthesis is a properly Berkeleian theory; however, such a reading was accepted in the eighteenth century.
- Then, to read *Siris* is nowadays an exercise of erudition: it is not possible to read it apart from a context on which it totally depends. The problem, now, is that we have forgotten this context – but this problem is ours, it does not mean that *Siris* is *in itself* a problem.

If the reading of *Siris* is a matter of erudition, it is nevertheless an important book for understanding Berkeley's philosophy and for gaining an interesting view of a philosophical discourse on the sciences in the middle of the eighteenth century.[39]

[38] Maybe the reader should not believe Berkeley when he writes: "it may, therefore, be pardoned if this rude essay doth, by insensible transitions, draw the reader into remote inquiries and speculations, that were not thought of either by him or by the author at first setting out" (*Siris*, § 297, *op. cit.*, vol. 5, p. 138).

[39] A first version of this paper was presented at the International Conference on Berkeley, organized by S. Parigi in Gaeta (Italy), 25–29 September 2007. I thank S. Parigi for her invitation, and the participants for their remarks, questions and encouragements – especially, R. Jakapi, M. Hight, G. Brykman and S. Charles. I also thank Camille Peterschmitt, Lawrence Principe and Steve Daniel who corrected earlier drafts.

Chapter 7
Berkeley and Newton on Gravity in *Siris*

Timo Airaksinen

Sir Isaac Newton speculates about aether, refraction of light, and the nature of gravity in his letters and in *Opticks* (1704). George Berkeley says in *Siris* (1744) that "[i]t is the opinion of Sir Isaac Newton" that aether may explain gravity (S #223).[1] According to Newton, gravity is created by the variable density of aether near and far from physical bodies. Berkeley recognizes this hypothesis. He asks, "Should not therefore gravity seem the original property and first supposed?" (S #225), and then he rejects this view as an illegitimate abstraction. In the end, he seems to say that gravity remains a mystery to the mechanistic science (*S* #227). His own philosophy of science cannot explain it, Newton cannot do it, and Descartes is wrong.

I review Berkeley's relevant views in detail and show how we can understand his explanations of the mutual attraction of bodies, gravity, and the related phenomena in natural philosophy. Berkeley was interested in natural science to the end of his career. I want to argue that Berkeley was a Newtonian natural philosopher in the sense that he knew Newton's work, learned from his speculations, followed his examples, and provided alternative models and explanations. He was critical of Newton, but he still works within the Newtonian framework of thinking. *Siris* owns much to *Opticks*.

7.1 I

In *Siris*, Berkeley discusses Newton's physics fairly and informatively, showing no hostility against it. It is clear that Berkeley was no Hutchinsonian enthusiast or a religious fundamentalist, as we say today. In his letter (vol. 8, #145) to Percival

---

[1] I refer to Berkeley's works by volume number and page or numbered paragraph, or by letter number: *The Works of George Berkeley, Bishop of Cloyne*, eds. T. E. Jessop and A. A. Luce, London, Thomas Nelson, 1948–1957. *Siris* is part of vol. 5, ed. Jessop (abbr. S).

T. Airaksinen (✉)
Susitie 10 C 32, 00800 Helsinki, Finland
e-mail: timo.airaksinen@helsinki.fi

S. Parigi (ed.), *George Berkeley: Religion and Science in the Age of Enlightenment*, International Archives of the History of Ideas/Archives internationales d'histoire des idées 201,
DOI 10.1007/978-90-481-9243-4_7,

from his Whitehall farm in Newport, Rhode Island he condemns it, as he does, throughout his works. He writes to Samuel Johnson (vol. 8, letter #260):

> As for Mr. Hutchinson's writings, I am not acquainted with them. I live in a remote corner, where many modern things escape me. Only this I can say, that I have observed that author to be mentioned as an enthusiast, which gave me no prepossession in his favour.

It is regrettable that some historians of science like Patricia Fara want to make him an enthusiast, or at least their sympathizer using, as evidence, Berkeley's criticism of Newton's mathematics.[2] Obviously, many of his contemporaries thought that Bishop Berkeley rejected Newton, but this contemporary view was not worth repeating. Berkeley was, indeed, critical of mechanistic physical theories, but he was also informed of them, respected them, and wanted to contribute to the corpuscular and mechanistic science of his own day. However, he never thought or said that the Bible was a natural science textbook like John Hutchinson's famous *Moses's Principia* (1724) argues.

Berkeley says he never read Hutchinson's book, but he praises Newton highly in *Siris* (S #245), as well as his geometry which is a crucially important point. Newton's mathematics was geometrical in nature. From the humble beginning in his *Notebooks* through the *Three Dialogues* and *De Motu* to the grandiose end in *Siris,* Berkeley discusses Newton's scientific ideas, for instance, on gravitation and understands them as deep and important problems not to be dismissed lightly. Even when Berkeley says that real causal explanations must be in terms of the human and divine will or agent causality and God's powers, he does not refer to the Bible. Biblical exegesis was of no philosophical interest to Berkeley, and this was the case even in the *Alciphron.* Enthusiasts depend on the Bible, but they do not trust reason, unlike Berkeley, so this is why he was not an enthusiast.

More than Newton, Berkeley opposes Descartes (S #232) and his famous aether vortex theory and his scientific explanations on that speculative basis, although Berkeley never bothers to discuss them in a thorough manner.[3] However, he gives a brief but accurate description of Descartes' mechanistic explanation of gravity (S #246) and he remarks that Newton "seems to have adopted somewhat not altogether foreign from this notion."[4] This is true: Newton's speculative and hypothetical

---

[2] Patricia Fara, *Newton, The Making of Genius,* London, Picador, 2002, pp. 103 ff. See also her contribution to *The Cambridge History of Science,* Vol. 4, Eighteenth Century Science, Cambridge, CUP, 2003, pp. 503–505.

[3] See Stephen Gaukroger, *Descartes' System of Natural Philosophy,* Cambridge, CUP, 2002, and *Descartes: The World and Other Writings,* ed. S. Gaukroger, Cambridge, CUP, 1998.

[4] See Gaukroger, *Descartes' System of Natural Philosophy,* pp. 165–166. Descartes, *The Principles of Philosophy* (1644, 4:23) explains gravity as follows. Celestial matter is in circular motion together with the Earth, so that the Earth is at rest in relation to it. The centrifugal force (or the force which is not required to turn the Earth around) makes the corpuscles of the celestial matter to move up in relation to the surface of the earth, but because they are located in plenum they cannot actually move. Large objects which contain a smaller amount of celestial matter in their pores than air are influenced less by the circular motion than the small particles of air. The celestial matter or the second element endeavors to move up. Thus, the small objects tend to replace the large objects, so that the large objects are driven down towards the center of the Earth. In other words, only if the large objects move down, the small objects can move up. All this is a function of larger amounts of celestial matter in the air. Air is light. Large objects are said to have more weight, which is the same as their gravity.

account of gravity resembles that of Descartes.[5] This is to be expected since Descartes played such a major role in the philosophy of nature. Berkeley, himself, makes many critical points in *Siris,* but he still works within the best scientific tradition of his day whenever he discusses scientific matters. Of course, *Siris* is not merely a scientific tract. Its author read and commented on a large number of ancient authors who were supposed to be so much nearer in time to the creation than us that their views are, both, more accurate and more reliable. This was a common view, and Newton held it too. In 1744 Berkeley was certainly not alone with this view, but at the same time, he was certainly among the most conservative thinkers in this semi-theological and backward-looking intellectual game. In Cloyne, some new things and trends may have escaped the good bishop, as he himself admits.

Ultimately, it does not matter much if we cannot explain everything; it is enough that we know that the course of nature is rational and well planned. We do not know *how* nature is to be explained, but we know *that* it is. Berkeley admits that all of the mechanistic and corpuscular explanations of gravity, along with many more facts, are doomed. We do not know how the correct explanations work. Yet, we know that by referring to God's will it is possible to know that they are explained. Berkeley's final position is, at the same time, that of epistemic resignation and a glorification of God's knowledge. We will return to this set of problems.

Everything is known by Him, but we can never reach that same level. Nevertheless, he never lapses in enthusiasm. He does not compare the status of science to the Bible. He does not say that science explains those newly found phenomena, although the Bible can since God can. God could tell us the truth, via the Book of Nature, if he wanted to do so, but obviously he does not want to since we do not know. Then we need to confess our failure. Why would God tell us about the new science in the Bible if he was unwilling to do so via Nature and its study? He already gave us knowledge in that way.

7.2 II

When Berkeley discusses gravity and the related phenomena like physical attraction and repulsion, he reaches the breaking point of his own philosophy of science, that is, the empirical theory of the laws of nature and their pseudo-causal role in explanations. Instead, he tries mechanistic explanations in terms of the corpuscular light/fire, something he was willing to neglect in the *Principles* (vol. 2 #50). It is remarkable that he does so. He realizes, and frankly admits, however, that he is unable to give an account of some key physical and biological phenomena (S #237) except by referring to God and agent causality, which is nothing new to his readers. Newton does the same, especially in his unpublished papers, and Berkeley knows it on the basis of his reading of

[5] On gravity, see also R. J. Boscowich, *A Theory of Natural Philosophy* (1763), tr. J. M. Child, Cambridge, Ma, MIT Press, 1966, pp. 144 ff. His point against aether is that it retards motion, a problem which worried Newton as well; see his *Opticks,* Amherst N.Y., Prometheus Books, 1730/2003, Qu 22. Newton needed aether to explain gravity, but his aether threatened to retard planetary motion, which is not supported by the facts.

Opticks.[6] He says that, "Newton himself, attests and subscribes" to the view that gravity depends on "the immediate action of an intelligent incorporeal being" (S #246). In addition, he remarks sarcastically that Newton does not always remember what his own view is. Of course, Newton could have said the same about Berkeley.

However, Berkeley and Newton seem to agree more than one might expect. Both of them, first, rely on the clearly formulated laws of nature, then offer their mechanistic hypotheses and finally rely on God's "immediate action."[7] Many why-questions look inexplicable to them, because they look like questions which can never be answered. Thus, the problems are not true empirical difficulties but genuinely impossible aporias. The hidden mechanisms which once promised to explain the ultimate mysteries of nature now look impotent and even worse – they look like mere speculations devoid of any reality. Hence, the legitimate problems of physics, biology, and physiology, as well, are to be formulated in terms of how-questions simply because they can be answered in terms of natural laws. Therefore, natural laws can be formulated empirically without escaping into something like an invisible or hidden mechanism. Once the laws are formulated, they can be tested and used with great success, as we already know. They are simple formulae which hide nothing and presuppose no hidden or occult entities. The success of science depends on them. Moreover, theoretical terms such as gravity can be defined in terms of these laws. Gravity is what the gravitational laws talk about. There is no mystery here, unlike in the case where the elastic aether should be used to describe an explanatory mechanism. And, why talk about hidden mechanisms when God can be referred to as the ultimate explanans?

The main difference between Newton and Berkeley, here, is that Newton kept his speculations mainly private by not publishing his relevant papers, whereas Berkeley published his similar thoughts in *Siris.* Hence, its contents are not so strange; what is strange is that Berkeley published his thoughts. And we must remember that *Siris* was a very successful book which sold well and was quickly translated, for instance, into French and Swedish. Its readers may not have found its speculations conspicuously unconventional, but they were certainly interested in its medical ideas, namely tar water as a panacea. It is also clear that *Siris* was not anything like a fully original book. Many of its ideas and speculations were already faintly familiar to its learned contemporary readers, for instance, its ruminations about the various roles of light and fire in the world. Of course, the details of *Siris* may have contained many surprises, but the outline of the work was traditional including its many, rather chaotic, references to Platonic and Stoic authors.

---

[6] B. J. T. Dobbs, *The Janus Faces of Genius: The Role of Alchemy in Newton's Thought,* Cambridge, CUP, 1991, reviews Newton's unpublished work very carefully.

[7] *Opticks* ultimately relies on God who "is more able by his Will to move the Bodies ... than we are by our Will" (Qu 31). Berkeley's idea of agent causality can be found in Newton as well. The true efficient cause is God. Notice that Newton mentions the human will as a cause, just as Berkeley does. Dobbs in her *The Janus Faces of Genius* emphasizes the role of God in Newton's pseudo-explanatory contexts.

7.3 III

In an interesting manner, biology or the philosophy of living nature becomes connected to physics in *Siris* at the explanatory level via attraction and repulsion. Fermentation is one of the key concepts already emphasized, also, by the alchemists.[8] Berkeley writes: "The phenomena of light, animal spirit, muscular motion, fermentation, vegetation, and other natural operations, seem to require nothing more than the intellectual and artificial fire of Heraclitus, Hippocrates, the Stoics (...), and other ancients" (S #277). He also writes in a manner which is not directly connected to what he said in the quotation above:

> And when it is said that all the motions and changes in the great world arise from attraction – the elasticity of the air, the motion of water, the descent of heavy and the ascent of light bodies, being all ascribed to the same principle; when from insensible attractions of most minute particles at the smallest distance are derived cohesion, dissolution, coagulation, animal secretion, *fermentation*, and all chemical operations; and when it is said that without such principles there never would have been any motion in the world, and without the continuance thereof all motion would cease; in all this we know or understand no more than that bodies are moved according to a certain order, and that they do not move themselves. (S #336, my italics)

Perhaps someone asks how the orderly motion of bodies explains fermentation. Anyway, biology and physics seem to be two disconnected and, in many ways, mutually independent realms, but when Berkeley must finally, at the end of his philosophical career, face such explanatory challenges as those posed by gravity, attraction, magnetism, and electricity, he admits that they are also relevant to biology. Hence, both physics and biology stay beyond the reach of Berkeley's own simple principles of explanation in terms of natural laws and are understood as nothing but observed regularities between ideas. The new phenomena look strange, various, and complex, but his old methods are simple. For instance, his explanations in terms of natural laws refer to observed ideas which are connected in a law-like manner to other ideas, but now the particles whose motions and effects are to be explained cannot be seen at all.[9] No ideas are at play here.

Berkeley expresses such a view of science very clearly indeed:

> Why may we not suppose certain idiosyncrasies, sympathies, oppositions, in the solids, or fluids, or animal spirit of a human body, with regard to the fine insensible parts of minerals or vegetables, impregnated by rays of light of different properties, not depending on the different size, figure, number, solidity, or weight of those particles, nor on the general laws

[8] Thomas S. Hall, *History of General Physiology*, Vol. 1, p. 207 says about van Helmont's theory: "The ferment is acting as an agent of 'Divine Light', that is of God." Here we see again a reference to God's actions and the role of light, but now in biology. *Siris* mentions van Helmont once (#49). Newton found fermentation to be an important term, which is to be expected as 'fermentation' has such a long history in alchemy (see *Dobbs, The Janus Faces of Genius*).

[9] G. Dawes Hicks, *Berkeley*, Bristol, Thoemmes Press, 1992/1932, is too optimistic when he says that according to Berkeley "attraction and repulsion turn out to be no more than tendencies or motions" according to the laws of nature (p. 215). Hicks dedicates a whole chapter of his book to *Siris* which shows that at that time it was actually supposed to be read.

> of motion, nor on the density or elasticity of a medium, but merely and altogether on the good pleasure of the Creator, in the original formation of things? From whence divers unaccountable and unforeseen motions may arise in the animal economy; from whence also various peculiar and specific virtues may be conceived to arise, residing in certain medicines, and not to be explained by mechanical principles. (S #239)

At least he is able to recognize the new frontiers of science. He does not seem to be too worried about his admitted explanatory failure because of his last recourse which is God's will and wisdom. Berkeley thinks that all these new and various mysteries of science make his faith stronger in the sense that God has created such wonderfully complicated things, and thus, He is able to control the world down to the last detail – this is a matter of His divine omnipotence. In an almost paradoxical manner, Berkeley welcomes the failure of the mechanistic explanations of Descartes and Newton even if it also meant the failure of his own corpuscular and mechanistic speculations. In *Siris*, he introduces them with such a feeling of excitement. Of course, all this supports his ultimate idea of explanation in terms of God's will and action. Yet, God is a mystery and so is his influence in the world, which is not to deny that God's will explains it all. He never tells the readers of *Siris* how God's will explains the phenomena. He only insists that it does. Berkeley writes in a mood that is, at the same time, pessimistic and optimistic. He is optimistic when he says that God's will explains. The pessimistic mood follows from the observation that such explanations are not connected to predictions. In his early philosophy of science, explanations and predictions belong together. This is why science is a practically useful art.

7.4 IV

The main explanatory problem and the ultimate unexplained mystery is gravitation or, more generally, the short and long distance forces of attraction and repulsion taken together with their biological counterparts. Fermentation is a representative example, as we already saw. Newton explains the gravitational mechanisms and mutual attraction of bodies by means of corpuscular aether and its variable density near and far from large bodies and inside their pores. Berkeley reviews Newton's *Opticks* and its explanation of gravity in terms of the aether hypothesis (Qu 21). A different explanation of attraction and repulsion is given by Newton in his early letter (1679) to Robert Boyle. This letter was published in 1744 which is also the year of the publication of *Siris*.[10] It is possible that Berkeley knew Newton's relevant letters, but it may be impossible to prove. Anyway, in his letter to Boyle, Newton says that "there is diffused through all places an aethereal substance [...], strongly elastic, and in a word much like air in all respects, but far more subtle." This is essentially the same account he gives in *Opticks* (Qu 18). However, he also says that "I do not know what this Aether is" (Qu. 21). It is just a hypothesis for him in his own near pejorative sense of the word. Newton's attitude is difficult

[10] See his letters in *Philosophical Writings*, ed. Andrew Janiak, Cambridge, CUP, 2004.

to understand: he condemns speculation and hypotheses, but then he is eager to present them. He writes countless unpublished pages of uninhibited speculation including ancient philosophy and hints at them in his *Opticks*. He says he does not know what aether is, although he tries all kinds of explanations by means of it.

Berkeley recognizes Newton's efforts, and after giving an accurate account of his view of gravity (S #223), he refutes it by showing that Newton's argument is viciously circular. I quote Berkeley; first his description of Newton's hypothetical explanation of gravity in terms of aether:

> And in passing from the celestial bodies to great distances, it [aether] is supposed to grow denser and denser continually, and thereby cause those great bodies to gravitate towards one another, and their respective parts towards their centres, everybody endeavouring to pass from the denser parts of the medium towards the rarer.

Next comes criticism:

> And gravity seems not an effect of the density and elasticity of aether, but rather to be produced by some other cause: [...] The elasticity of which fluid is supposed to depend upon, to be defined and measured by, its density; and this by the quantity of matter in one particle, multiplied by the number of particles contained in a given space; and the quantity of matter in any one particle or body of a given size to be determined by its gravity. (S #225)

Actually Newton talks about the rarity of aether both in the first part of his letter to Boyle and in *Opticks* (Qu 21), but this may well be taken to be the opposite of density. What Berkeley says is that aether is an elastic medium and elasticity logically entails density which, again, entails "quantity of matter" – but this cannot be determined without reference to its weight (mass) and gravity. Hence, Newton's account of gravity in terms of the density or rarity of aether is viciously circular. It is sometimes said that *Siris* is poetry and contains no arguments which is obviously not true, as we see here. Berkeley designs a nice logical argument against Newton.[11]

Newton may be guilty of another circularity, as well. He explains gravity in terms of the rarity of the elastic aether in space and a vacuum saying that bodies endeavor to reach a position in the area of lesser density of aether or move from a higher to lower pressure, if we want to put it that way. All of this action depends on aether being an elastic medium. Next, Newton tries to give a mechanistic explanation of elasticity. Thus, he writes: "And so if anyone should suppose that *Aether* (like our Air) may contain Particles which endeavor to recede from each other.... The exceeding smallness of its Particles may contribute the greatness of the force by which those Particles may recede from one another" (Qu 21). The mutual repulsion of the particles of aether tends to create a vacuum between them and, thus, an environment where matter is rarer. This looks circular because elasticity is required to explain gravity, and now elasticity itself is explained in terms of a repulsive force which pushes the small particles of aether away from each other. This is not circular, of course, if the explanations of gravity and repulsion are different so that gravity is explained by A and repulsion by B, when A and B are now two fundamentally different explanantia. This is also to say that gravity

[11] See my 'The Path of Fire: The Meaning and Interpretation of Berkeley's *Siris*', in *New Interpretations of Berkeley's Thought*, ed. Stephen H. Daniel, Amherst, NY, Prometheus/ Humanity Books, 2008, on the various types of philosophical responses to *Siris*.

and attraction must be two different natural phenomena. Yet, when one reads *Opticks* and Newton's relevant letters, one does not find two different explanations. On the contrary, gravity and attraction are both explained in the same way in terms of the properties of this occult medium, aether (see below). Yet, a careful reader may sense some hesitation here. Gravity works between large bodies and long distances, unlike attraction and repulsion. Moreover, gravity is not associated with any counter-force in the same way attraction is associated with repulsion.

Berkeley presents an additional and almost rhetorical argument against Newton: "The phenomena of electrical bodies, the laws and variations of magnetism and, not to mention other kinds, even gravity, is not explained by elasticity, a phenomenon not less obscure than itself." (S #243). Berkeley certainly is right when he pays attention to the relation between gravity and elasticity. *Siris* contains some apt philosophy of science.

Next, Berkeley considers an alternative account of gravity and dismisses it quickly on familiar grounds. He does not accept gravity as an abstraction which is inadmissible in a context where we speak of real and existing things. However, he also condemns fictional spiritual forces independently of whether they are abstractions or not. In other words, he does not reject gravity only because it is an abstraction but also because it is a "spiritual force" which is independent nonsense. He writes:

> Should not therefore gravity seem the original property and first supposed? On the other hand, if force be considered as prescinded from gravity and matter, and as existing only in points or centres, what can this amount to but an abstract, spiritual, incorporeal force? (S #225)

In fact, both Newton and Descartes share this same view with Berkeley. Gravity and the gravitational force are not an original property which belongs to all matter. For instance, Descartes' matter is nothing but pure extension, *res extensa,* and thus, matter cannot be associated with such an original gravitational force. Berkeley, however, has already listed gravity as one of the (illusory) primary qualities of matter in his *Three Dialogues* (vol. 2, p. 187): "For the clearer understanding of this, you must know sensible qualities are by philosophers divided into primary and secondary. The former are extension, figure, solidity, gravity, motion, and rest." Here, gravity is, by implication, called a sensible quality when it is also called an *occult* quality in *De Motu* (vol. 4, p. 32): "But since the cause of the fall of heavy bodies is unseen and unknown, gravity in that usage cannot properly be styled a sensible quality. It is, therefore, an occult quality."[12] Here, Berkeley's interpretation of the concept 'occult'[13]

---

[12] Bertil Belfrage translates the same text as follows: "But since strictly speaking the cause of the fall of heavy bodies is unseen and unknown: gravity in that sense cannot be properly called a sensible quality..." This translation makes the *Three Dialogues* and *De Motu* look mutually less inconsistent.

[13] What Berkeley means by 'occult' is a key question. Berkeley's use of 'occult' is ambiguous with respect to its Aristotelian sense. In the *Principles* (vol. 2, #102) Berkeley draws a distinction between occult qualities and the qualities of insensible particles. In *De Motu* (vol. 4, p. 32) this view is no longer valid. Now, what is not sensible is, therefore, occult which makes force and gravity occult things. In *Siris* both ideas seem to be used. For example, in #175 an "occult universal nature" is mentioned. He mentions occult when he discusses ancient philosophy, and thus, it is not clear what his own view is. But he still says that gravity is an occult property (S ##239–240) even if he denies

is devoid of its normal, semi-mystical flavor mentioned, for instance, by Newton in *Opticks* (Qu 31): "And the Aristotelians gave the name of occult qualities, not to manifest qualities, but to such qualities only as they supposed to lie hid in bodies, and to be the unknown causes of manifest effects ..." According to Newton, gravity is not one of the occult qualities which "result from the specific forms of things." Berkeley's view seems to be that all unperceivable qualities must be called occult independently of any such Aristotelian forms of things. This is another example of how Berkeley comments on Newton, sometimes explicitly but often implicitly as well.

In his letter to Richard Bentley (1692) Newton, again, denies that gravity is "innate, inherent, and essential to matter," that is, an occult quality. The reason is that this implies action at a distance, which appears as impossibility to Newton. Berkeley and Newton agree here but for different reasons, both philosophical and scientific. Newton's gravitational hypothesis may require a mechanism such as the actions of aether particles, which makes gravity possible. Berkeley's attitude to explanations in terms of hidden mechanisms is negative but not wholly consistent (see for instance S #234). But, in that case how could he give any type of an account or explanation for gravity or attraction or magnetism or for the forces of electricity? In *Siris*, he tries to do just that. Sometimes in *Siris*, he plays with hidden mechanisms as if they would all be as self-evident as those well-known mechanisms which God created. Nature, indeed, contains many mechanisms:

> There are innumerable fine and curious parts in a vegetable body and a wonderful similitude or analogy between the mechanism of plants and animals. And perhaps some will think it not unreasonable to suppose the mechanism of plants more curious than even that of animals. (S #31)

But then, he also finds it necessary to reject those hidden and invisible mechanisms which are behind the phenomena called gravity, attraction, magnetism, and electricity. His reader struggles to form a consistent view of what the philosopher is actually doing here. It is clear that the human body is a complicated mechanism for a physiologist, but the ontological status of the gravitational mechanism must be something else. It is clear that the physiological mechanisms are not visible in the same sense as clockwork is. Therefore, at least three different types of mechanisms should be recognized with two of them being acceptable in science, although only one is visible.

---

that gravity is an "original property" (S #225). This is, indeed, confusing. He uses 'occult' both in the traditional and the empiricist way: what is occult may be a hidden nature of a thing which gives it its observed properties, or 'occult' means what we cannot come to observe in science. It is strange that Berkeley wants to use 'occult' at all because both meanings are beyond science. There are no hidden natures nor are there – within the scientific discourse – entities which science could not come to explain. It does not make sense to say that gravity exists, but we *know* that we can never explain it. The key idea is that of something being *hidden* in nature. What is occult is hidden and, therefore, also unobservable. But 'hidden' means much more than merely 'unobserved' or even 'unobservable.' – Is gravity an abstraction? I cannot deal in this paper with the problem of abstract terms and entities. Very briefly, abstract terms for Berkeley may be useful in science and this way also legitimate, but one should never think that what they refer to exists, in the strict sense of the term.

Before we go on, we need to notice a choice of words Berkeley makes when he discusses Newton's theory of aether. Newton, himself, compares aether with *air*, and he does so often and consistently, but Berkeley compares aether with *light* (S #223). In *Opticks* (Qu 28) Newton suggests that aether may be "very thin vapours ... arising from the atmospheres of the earth, planets, and comets ...." According to Berkeley, Newton says aether is to be more subtle than light, which is true. Yet, somehow Berkeley seems to miss Newton's point when he writes:

> It is the opinion of Sir Isaac Newton that somewhat unknown remains in vacuo, when the *air is exhausted*. This unknown medium he calls aether. He supposeth it to be more subtle in its nature, and more swift in its motion, than light, freely to pervade all bodies, and by its immense elasticity to be expanded throughout all the heavens. (S #223, my italics)

This little detail is not as innocent as it may initially look because Berkeley wants to promote his own pet theory of the celestial and occult light/fire which should replace aether in physics (see my appendix). Therefore, Berkeley does not need anything like Newton's aether or Descartes' second element and celestial matter. For Berkeley "aether is fire, and if fire, light" (S #227). Berkeley is not quite accurate when he describes Newton's views here. He wants to promote his own theory, and thus, he reads Newton in that new light, too. But then, we are now discussing *Siris* which is full of such intriguing little details.

Newton discusses attraction and repulsion, but not gravity, in the first part of his 1679 letter to Boyle. Here, gravity and attraction seem to be two different things. Gravity appears only at the end of the letter, and there his hypothesis is different from that in *Opticks* which Berkeley mentions in *Siris*. In the first part of his letter Newton discusses something else, for instance, what happens when two polished glass panels are pressed together or a fly walks on the surface of water. What happens when the glass surfaces are pressed together? First they tend to resist contact, but when a greater force is applied they collapse together so that they are difficult to separate. A fly walks on top of water without wetting his feet, as Newton says, which means that its mass is not great enough to press it down into the water. The initial repulsion dominates. This theory explains an aspect of attraction and repulsion. Berkeley worries about this same problem which he obviously inherits from Newton: "And when flies walk in water without wetting their feet, it is attributed to a repelling force or faculty in the fly's feet. But this is obscure, though the phenomenon be plain" (S #235). Once again, he is right.

Newton's 1679 account of attraction and repulsion is as follows. Aether is rarer or less dense inside of bodies than it is outside of bodies, and it is denser far from them and rarer near them.[14] Now, when bodies approach each other the density of aether diminishes between them, obviously because the two masses together bring about extra rarefaction of aether between them. Aether is rare near the bodies, and

[14] The same idea is repeated in *Opticks* (Qu 21). Newton did not change his ideas about this feature of aether.

now two of them together bring about extra rarefaction. This also causes resistance preventing the bodies from collapsing together: the rarefied aether endeavors to keep its former density. Next, something happens which is consonant with Berkeley's account and which may make us to think of gravity in this context. According to Newton:

> But at length, when they come so near together that the excess of pressure of the external aether which surrounds the bodies, above that of the rarefied aether, which is between them, is so great as to overcome the reluctance which the bodies have from being brought together, then will that excess of pressure drive them with violence together ...[15]

Newton's account of attraction and repulsion of bodies, obviously, presupposes that some additional force is applied to the bodies; otherwise the constant high pressure from above cannot overcome the ever increasing resistance brought about by the rarefaction of the aether between the bodies. The whole scenario implies an external source of force. The two examples of glass panes and Jesus flies are well taken here.

In the last section of his letter to Boyle, Newton explicitly refers to gravity. Now, he asks us to assume that the particles of aether are larger outside and smaller inside bodies. This is easy to accept because only small particles fit into the pores of the bodies. The particles of aether also become finer and finer when their distance from the Earth decreases. Now, when a body approaches the Earth, it comes into contact with the finer aether which is better adapted to lodge in its pores, and thus, the larger particles are driven out and replaced by the smaller and finer particles. But, what has this to do with gravity? Newton says that this replacement cannot take place "without the bodies descending to make room" for the smaller particles. Such an account is no longer mechanical but clearly teleological or neo-Aristotelian. The body endeavors to reach a position in which it can take in the smaller aether particles which fit its pores better than the larger ones. If it does not descend, the invading aether particles are of the wrong size. The body *must* descend in order to reach its perfect position, because the aether "will endeavor to get out and give way to the finer aether below." This is why the body descends to the Earth which appears to observers as the effect of gravity and suggests the existence of gravitational force.

At the end of his letter, Newton confesses that he has very little enthusiasm for such speculations as these. To his modern reader, it is interesting that when Newton drops his empiricist and mathematical mode of thought and allows himself to enter the shady world of hypotheses, he does not see anything wrong with active principles, teleological arguments, or the language of endeavor or conatus. This also illuminates Berkeley's new attitude toward science in *Siris* where he follows Newton more faithfully than ever.

[15] Newton plays with the romantic idea of violent and even catastrophic effects in various places in *Opticks* (Qu 31). Berkeley, once again, follows his example on *Siris* (##158 and 176).

7.5 V

Obviously, after reviewing and criticizing Newton, Berkeley must say something more positive about gravity and the other types of the forces of attraction. He also recognizes the need to mention electricity and magnetism just like Newton does in his *Opticks* (Qu 31), and he goes on towards such biological phenomena as fermentation which is an active principle and the main source of heat and life in bodies. Newton discusses fermentation in *Opticks* (Qu 31) – again we see how closely Berkeley follows Newton's example. All of these problems require a description in terms of natural laws and may be open to mechanistic explanations. Perhaps, they are all dependent on the forces of attraction and repulsion. Berkeley proceeds in a threefold manner without being able to unify his account in the end. He seems to try all of the methods of explanation just to see how they work. First, Berkeley gives a pseudo-Newtonian hypothetical mechanistic account of gravity in terms of his own light/fire. Next, he returns to the *De Motu* -style of an empirical account of the laws of nature, including those which describe the properties of gravity, and ultimately, he must recognize the futility of trying, somehow, to give an account of all the various types of natural phenomena within one scientific explanatory framework, whatever it is. He maintains, of course, that he can do it by referring to God's arbitrary will, but then he steps out of the world of the modern empirical science and its future development.

First, Berkeley discusses, in *Siris,* the effects of the material substance he calls light/fire which takes the place of Newton's thin air-like aether. Light/fire is the same thing as the Newtonian material light as a hail of small bullets, but to this Berkeley adds the rather archaic idea of fire.[16] Light is often hot, too, so their empirical connection is easy to see. To support his own case, he says that his own light/fire is not homogenous like Newton's aether, and therefore, it can, in principle, explain more. Light's many properties can be seen in its spectrum created by a prism, as Newton himself has shown. Strangely enough, Berkeley calls light/fire aether, too. This verbal play has confused many commentators of *Siris.* He writes:

> seminal principles have their natural existence in the light, a medium consisting of heterogeneous parts, differing from each other in divers qualities that appear to sense, and not improbably having many original properties, attractions, repulsions, and motions, the laws and natures whereof are indiscernible to us otherwise than in their remote effects. And this animated heterogeneous fire should seem a more adequate cause, whereby to explain the phenomena of nature, than one uniform aethereal medium. (S #229)

In this quotation some particles of light/fire have attraction and repulsion among their "original properties" which is an intriguing statement when we remember

---

[16] Fara says that Berkeley's light/fire is spiritual, which is a clear mistake in the present context. See my 'The Path of Fire'. She also says that *Siris* is a "long tract," which is not quite accurate. *Siris* has some 130 odd pages (Fara, *Newton,* pp. 103 and 107).

what he says in the *Three Dialogues* about gravity as a primary quality of the objects of perception. What was a primary quality has now been changed into an occult entity when 'occult' is understood in its Aristotelian sense. All original properties are occult (but see my n. 13). We need to take this last development as one of Berkeley's many thought experiments in *Siris*. The problem is, obviously, that he never tells us that it is such. And then, it is difficult to see what his main point is. Moreover, he never tells us exactly how light/fire is supposed to explain gravity and other forces of attraction. He just says that light/fire might do the work without describing any hidden mechanisms as Newton and Descartes do. The point is that God controls light/fire which then brings about the gravitational phenomena.

Notice that, in spite of what Berkeley says, Newton is ready to accept the heterogeneity of his aether, too. Some of its particles may be larger than others (Letter to Boyle), so why not postulate differences in other dimensions as well? Air and atmosphere consist of heterogeneous elements which is that "ambient heterogeneous fluid called air," as Berkeley admits in *Siris* (S #33), and this means that Newton's aether may be heterogeneous as well. But the idea of aether is, for Newton, quite tentative anyway, so it is difficult to tell. Berkeley wants to take the alleged homogeneity of Newton's aether seriously. Later, he seems in a rather enigmatic manner to contradict his earlier promotion of a heterogeneous explanans when he writes as follows:

> Nature seems better known and explained by attractions and repulsions than by those other mechanical principles of size, figure, and the like; that is, by Sir Isaac Newton, than Descartes. And natural philosophers excel, as they are more or less acquainted with the laws and methods observed by the Author of nature. (S #243)

We learn here that Newton is a better natural philosopher than Descartes. But, Berkeley has already said that the mechanical principles do not explain gravity in such a way as Newton's explanation in terms of aether and its endeavors. Now, he says that attractions and repulsions may explain something, although he has also said that they are an *explanandum* and not an *explanans*. Then he refers to his own light/fire as a tool of devising mechanical explanations which might do the work. Moreover, he says that his light/fire is heterogeneous, unlike Newton's aether. In this way, Berkeley wants to return to mechanical explanations by means of heterogeneous material entities. Thus, the quotation above presents or exhibits several problems of interpretation: first, how can attraction explain when it should be explained; second, why should we lean on mechanisms; and third, why condemn heterogeneity of "size, figure, and the like" when light/fire is heterogeneous and, therefore, works so well in explanations? A simple solution is, of course, that Berkeley wants to promote Newton's laws of gravitation. Once again, he thinks that successful science does not need anything else but the natural laws.

Berkeley's own method, which for him is the best of them all, refers primarily to the will of God and then to the laws of nature (which are mere observed regularities). Yet, his notion of light/fire does not disappear, because it has its central metaphysical role throughout *Siris*. Light/fire mediates between God and the physical world. God's will is realized by means of light/fire: "the inferior instrumental cause is pure aether, fire, or the substance of light (...), which is applied and determined by an Infinite Mind in the macrocosm or universe" (S #154). Thus, scientific explanations in terms of light/fire are unique to *Siris* and should be contrasted to the more familiar Berkeleian language of the laws of nature as we find them, for instance, in *De Motu.* Then, Berkeley quickly returns to his earlier view according to which attraction and repulsion are (legitimate) abstract linguistic terms taking the role of theoretical terms as follows:

> The laws of attraction and repulsion are to be regarded as laws of motion; and these only as rules or methods observed in the productions of natural effects, the efficient and final causes whereof are not of mechanical consideration. Certainly, if the explaining a phenomenon be to assign its proper efficient and final cause, it should seem the mechanical philosophers never explained anything.... (S #231)

Newton talks, for instance, about the endeavor of the aether to return to its "natural state of condensation" (his letter to Boyle), and Berkeley is, understandably, quite opposed to this kind of teleological Aristotelian language and its rhetoric of final causes. He has, himself, suggested a perfectly mechanistic efficient cause and explanation of gravity in terms of the heterogeneous matter called light/fire even if he also mentioned something called active principles.[17] Berkeley wavers between his old *De Motu* and new *Siris*, between the laws of nature and the promising new idea of light/fire, and between the rejection and re-establishment of mechanism as an explanans. In a strange way, the laws of nature are not enough, and the new mechanisms are too much: gravity seems to be more than a mere law-like description of how material objects move, and he has firmly rejected mechanisms a long time ago by means of cogent arguments. But worst of all, such old terms as occult entities, endeavors, active principles, and original properties refuse to disappear from the Newtonian language Berkeley so much admires.

Strangely enough, Berkeley suggests a mechanistic explanation by means of the variable inherent properties of light/fire. The solutions to this riddle may be as follows: the only possible account of causality is in terms of God's will (S #237), as Berkeley always emphasizes; light/fire is God's special instrument of influencing the physical world. Therefore, light/fire-explanations work where those in

[17] In *Opticks* (Qu 31) Newton introduces two "active principles," the cause of gravity and the cause of fermentation, which increase motion which is "always decreasing" in the world. This is to say that aether is an active principle. Alternatively, the explanation is God who is "able by his Will to move" the Bodies"? (Qu 31). Berkeley mentions "active principles" in *Siris*, e.g. #68. He has already said: "A little attention will make it plain to any one, that to have an idea which shall be like that active principle of motion and change of ideas, is absolutely impossible" (*Principles* #27). See also *De Motu* (#31). In the *Dialogues* Philonous makes God an active principle: "How often must I repeat, that I know or am conscious of my own being; and that I myself am not my ideas, but somewhat else, a thinking active principle that perceives, knows, wills, and operates about ideas" (#223). The same point is made in *Alciphron* (vol. 3, p. 290). But in *Siris* he mentions "fire, that diffused and active principle" (#158). See also Lisa Downing, 'Berkeley's Natural Philosophy and Philosophy of Science', in K. P. Winkler, ed., *The Cambridge Companion to Berkeley*, Cambridge: CUP, 2005, p. 254.

terms of aether must fail. The light/fire is like a screwdriver that the watchmaker uses to change and correct some of the details of his ingenious design in the material world. The problems of interpretation are immense; therefore I cannot go into details here. To put it briefly, light/fire does not explain mechanistically, although light/fire consists of real particles just like Newton's aether. Somehow, light/fire is *directly* connected to God's will – in this sense it is active – and to the idea of agent causality as the only genuine type of causality. This is why it can be used in physics to give an explanation of gravity. Light/fire does not explain merely because it constitutes a new idea of mechanism, on the contrary, it explains in spite of the fact that it looks like a mechanism – it is the true, unadulterated causal vehicle or the main instrument of God's will. This is one of Berkeley's main ideas in *Siris* and, perhaps, the one which is original to him, although his full account of it all remains sadly lacking. Berkeley is unable to say much about it. How light/fire actually works as a causal factor in natural philosophy requires a separate study which cannot be done in this paper. We should ask, for instance, if X is light/fire which causes Y, what is Y? How does X bring about Y? How does God's will change the actions of X on Y?

7.6 VI

Next, we discuss the riddle of those motions which are too varied and specific to allow for systematic observation. Here, Berkeley comes close to admitting his scientific defeat, as he must admit that the phenomena unraveled by modern science are so varied and complicated that they may ultimately be beyond our human comprehension.[18] Of course, God understands them all and commands them by using light/fire, but this is only a small consolation to an eager enquirer and, perhaps also, to Berkeley himself. The most worrying thing from his point of view is, certainly, the fact that the benevolent usefulness of the familiar laws of nature, which are so important in the *Principles*, seems to be lost from the world of new science, as it is described in *Siris*. If the laws of gravity, magnetism, electricity, and the associated biological phenomena like fermentation are so diverse and various, as he emphasizes (S #235), that they cannot be accounted for in science except by referring to God's will and actions, these phenomena cannot be predicted. Therefore, the usefulness of science is lost, because it all depends on our ability to make predictions.

[18] In his questions to *Opticks* Newton mentions a large number of different problems. The physical world now appears to be very complex. In *Siris* Berkeley adopts the same view and he uses the same words in the same order as Newton when he speaks of the new problems of science, that is, when he says that "The laws of gravity, magnetism, and electricity are diverse" (S #235, cf. the first paragraphs of Qu 31). However, the ancient thinkers were able to formulate them (S #277) by means of their theory of fire which is also light. What Newton has shown to be a world of great complexity and variety, the ancients have already explained. In addition, Newton thought so (see Dobbs, *The Janus Faces of Genius*). Berkeley follows Newton quite faithfully.

In sum: Two different reasons for unpredictability exist: first, the explanations by light/fire do not entail predictions, and second, some natural phenomena are so varied and specific that no law-like propositions apply to them. Let us now focus on the second reason.

We observe a phenomenon, and we know what follows from it. As Philonous says: "for by observing and reasoning upon the connexion of ideas, they discover the laws and methods of Nature, which is a part of knowledge both useful and entertaining" (vol. 2, p. 243). Now, systematic observations fail and then the problem is that there is no way of predicting what God will do in such cases. When the phenomena and their laws are really diverse and various so that their "mechanistic laws" are not available to us, then only God can help, but then, we cannot predict. We can only predict if we have those laws available to us. Especially in biology, many phenomena are mysterious in this sense. For instance, how tar water works on the animal body and mind cannot be explained merely in terms of perceived motions and their law-like connections, and *Siris* is all about tar as a medicine:

> whence divers unaccountable and unforeseen motions may arise in the animal economy; from whence also various peculiar and specific virtues may be conceived to arise, residing in certain medicines, and not to be explained by mechanical principles. For although the general known laws of motion are to be deemed mechanical, yet peculiar motions of the insensible parts, and peculiar properties depending thereon, are occult and specific. (S #239)[19]

Berkeley thinks that a category called occult and specific motions exists, that is, they cannot be perceived (occult) and their nature cannot be captured by general scientific laws (specific). What is varied and specific is not law-like, as I read it.[20] Again, he follows Newton who writes as follows: "I [would] rather infer from their cohesion that their particles attract one another by some force, which in immediate contact is exceedingly strong, at small distances performs the chymical operations above mentioned" (*Opticks*, Qu 31). Newton lists an impressively large number of chemical phenomena which can hardly be subsumed under a few simple laws. He makes a contact between the forces of physics and chemistry in a way which is, at the same time, similar to and different from Berkeley's account. Newton applies one familiar force, attraction, to physics and chemistry, although this looks like a

---

[19] Notice that Berkeley speaks about mechanical laws and contrasts them with the laws which are varied and specific. His point may be that the first type of laws could, in principle, be accounted for by hidden mechanisms. The second types of laws are different in this respect.

[20] Berkeley's notion of an occult quality seems to be simpler than Newton's (Qu 31) who refers to its Aristotelian Medieval origin. Berkeley seems to mean by occult something which cannot be perceived, or an unobservable quality. However, light/fire cannot be perceived, and thus it is called occult (see Appendix). Because Berkeley accepts light/fire which he calls occult, he must accept in *Siris* occult qualities too. On the other hand, it is possible to find traces of the Aristotelian/Newtonian usage of 'occult' in the *Principles* (#102) where occult and insensible things are not identified and in *Siris* (#175) where he mentions an "occult universal nature."

far too simple solution. Berkeley allows a wide diversification of forces and motions which become "occult and specific." It is, indeed, as if Berkeley hesitated because of all the explanatory problems which he now sees facing his original, basic account of science in terms of the empirical laws of nature. Too many and diverse fields of phenomena have come to sight, as *Opticks* shows, and one single force like attraction does not seem to be enough. Many more specific forces are needed, but then, they remain hidden to the philosopher.

All of this is fine when we discuss biology and, for instance, fermentation or the healing effects of tar water and other medicines – but is gravity to be classified, also, as a non-mechanistic feature of physical bodies? Berkeley comes very close to saying this. Yet, he knows that the gravitational laws can be formulated, and that gravity can be understood on that basis. He does not pay much attention to magnetism and electricity, although as we already know he says that "[t]he laws of gravity, magnetism, and electricity are diverse." (S #235). He says; "The laws of these motions are various." (S #236). How can there be so many laws of gravity? But Newton, himself, is ambiguous about the mutual relationship between the laws of attraction and gravity. For some reason these phenomena are so strange and unpredictable, one law is not enough. This is supposed to mean that no unifying, single, and simple description of these motions can be found which logically entails a lack of natural laws and the impossibility of scientific prediction. Berkeley clearly believes that gravitational laws are available to science, but he stops there. Electricity and magnetism will prove to be more difficult to master, and physiology is an even more complex field of science.

7.7 VII

Notice how Berkeley's early views of the natural law – based pseudo-causality have changed. A natural law is, typically, established in a situation in which a perceiver first perceives a phenomenon A which is then regularly followed by B. This is the main idea. He can then predict, reliably, that B follows after A. A and B are, thus, connected in a law-like fashion which corresponds to the psychological expectations of the observer. Next, suppose you perceive a magnet which is approached by another magnet so that they attract each other and collapse together. Of course, in some cases the same magnets repel each other (if their ± pole configuration is changed). Yet, the visual perception of the magnets cannot give you a hint of what is going to happen. The relevant ideas remain unchanged. You can visually observe as many magnets as you like, and yet, you cannot predict whether a new brace of magnets will repel or attract each other. Of course, once you try, you may place the properly marked magnets so that they repel or attract each other. Then, the rule and the law of attraction and repulsion is easy to apply. At first, however, one cannot visually predict what will happen. One must try and see how the respective magnetic poles are arranged in any given situation. In other words, the visual cues to the situation become void, and no lawlike connection can be established. Such cases are all diverse, in Berkeley's sense of the word.

The same problem applies to other types of attraction as well, including gravity, as Berkeley says. Gravity, too, is an occult phenomenon. If the laws of nature connect two perceptions, as they should, or they are God's benevolent words in the Language of Nature, gravity may also prove to be an occult and specific motion. The crucial question is: what is the first idea A which is then followed in the law-like fashion by another idea B? Many different A's can be suggested, and thus, there may be many laws of gravity. For instance, when A is "a brick on the roof" no visible event description applies in the situation. What we have here is a dispositional property of the brick being heavy and prone to fall down. Such dispositional properties became a real problem to the Logical Empiricists of the early twentieth century. Berkeley says, "all the phenomena in nature are produced by motion" (S #234), but this mechanistic thesis is obviously a far too simple view.

This problem is discussed already in *De Motu*:

> But since the cause of the fall of heavy bodies is unseen and unknown, gravity in that usage cannot properly be styled a sensible quality. It is, therefore, an occult quality. But what an occult quality is, or how any quality can act or do anything, we can scarcely conceive – indeed we cannot conceive. And so men would do better to let the occult quality go, and attend only to the sensible effects. Abstract terms (however useful they may be in argument) should be discarded in meditation, and the mind should be fixed on the particular and the concrete, that is, on the things themselves (vol. 4, p. 32).

Here, gravity is called an occult quality, but this is not to say that gravitational motion, itself, is occult. Therefore, Berkeley speaks about occultism in two different contexts: there are occult motions in biology and chemistry (*Siris*) and occult qualities in physics (*De Motu*).

However, even if we can see physical bodies falling towards the center of the Earth, we may not describe them in terms of the required perceptional A-B-pattern. However, Berkeley utilizes two different concepts of natural laws: one is according to the A-B-pattern and the other means any observed regularity. The latter one means a regular motion as such; the former is somehow connected to the analysis of causality since A is popularly – even if mistakenly – called a 'cause of B'. In the latter case, we have the Newtonian laws of gravity which accurately describe the fall of a brick but which cannot indicate any cause of the fall. This is the crucial difference: in one case there is an apparent cause and a law, and in the other case there is no visible cause, but yet, there is a law. Here, Berkeley's philosophy of science appears to lose its contact to his early A-B idea of physical pseudo-causality which he formulated in terms of observed causes and effects, and consequently, to that aim of science which is to describe any regular and perceived motions there are.

In sum: Berkeley recognizes at least four different cases when he discusses causality. First, a simple psychological *expectation* that an idea B which will always follow another idea A makes A look like the cause of B. Second, the *law-like* structure of events, itself, suggests hidden causality. Originally A and B were supposed to be perceptual ideas, but this condition starts looking too narrow very soon. A may not be visible at all. An example is the laws of gravity which are causal laws. *Siris* speaks about mechanical laws in this context when causal laws are meant. In this sense, when object A falls down, we explain it in terms of Newton's law of gravity, and we think that this is a causal explanation. The third type of causality is a *mechanism* which explains an event. Some mechanisms are invisible, some are hidden, and some are visible. Here is a good example of a failed explanation in terms of a hidden mechanism:

> Homberg, nevertheless, holds in general that acids are shaped like daggers, and alkalies like sheaths, and that, moving in the same liquor, the daggers run into the sheaths fitted to receive them with such violence as to raise that effervescence observed in the mixture of acids and alkalies. But it seems very difficult to conceive how or why the mere configuration of daggers and sheaths floating in the same liquor should cause the former to rush with such vehemence, and direct their points so aptly into the latter, any more than a parcel of spigots and fossets floating together in the same water should rush one into the other. (S #132)

Aether plays the crucial role of a mechanical explanans when it is used to explain gravitational motion. The motion of the particles of aether brings about those perceived motions we call the effects of gravity. When an apple falls to the ground, it is pushed down by some invisible corpuscles. As Berkeley may say, such explanations are useless and scientifically inadmissible, because we cannot perceive the cause. His own invisible light/fire is supposed to be an admissible mechanism, why? The reason is that it is so closely connected to God and his actions; it is as if it were God's messenger. In addition, *God* is a true cause (of the fourth type) according to Berkeley – and ultimately, Newton as well as Berkeley sometimes recognizes this. Alas, God and mechanisms share the main problem. Both are invisible and unpredictable. We need perceived law-like connections between phenomena, and we want well-formulated and accurate natural laws which predict well. They and they only are useful. In this sense, God is a mere luxury, some kind of a consolation to those who despair when they realize how difficult it is to formulate the various and diverse natural laws which govern causality, magnetism, electricity, and fermentation.

I do not think Newton's and Berkeley's philosophies of science differ much. Both authors want the same thing, the Newtonian laws of nature, while everything else is secondary. For Newton, himself, this means some strange hypotheses, but for Berkeley it means the whole of *Siris*. Sadly, this is to say that *Siris* was the last scientific and theological effort by Berkeley when he already had lost his hope that the pure logic of ideas could work. However, he never fully rejected his early views. In *Siris*, the reader meets constant references to immaterialism, as if Berkeley thought that the *Principles* and *Siris* could be seen to form a continuous unity. Once again, he states with great authority; "Natural phenomena are only natural appearances. They are, therefore, such as we see and perceive them" (S #292). Hence, in *Siris*, the bishop plays with his own "children of imagination grafted upon sense." Such is one of the many paradoxes of *Siris*.[21]

---

[21] I am truly grateful to Bertil Belfrage for his detailed critical comments, and also to David Berman. The first draft of this paper was read to the Department of History and Philosophy of Science, University of Cambridge, January 2006; my thanks are due to Martin Kusch. My research has been funded by grants from TEKES, the Academy of Finland, and the University of Helsinki.

7.8 Appendix

This is a list of terms used in *Siris* to characterize light/fire. The list is not complete, and it contains terms which are used in an ancient historical context and which are, therefore, suspicious:

sun's light S #32
solar fire 37
solar emanations 43
light 110
(fire is acid) 128
pure aether, fire, or spirit 150
aether, fire, or spirit 151
aether 151
aether, or pure invisible fire 152
pure spirit or invisible fire 157
pure fire, or the spirit of the universe 159
pure aether, or invisible fire 162
aethereal fire or light 164
aethereal substance or fire 166
pure fire, aether, or substance of light 169
tunicle of the soul 171
luciform aethereal vehicle 171
celestial aether 181
pure elementary invisible fire 190
real actual fire 198
fine aethereal fire 199
(fire is sulphur 202)
pure flame 203
occult fire or spirit 210
occult light 210
aethereal fiery spirit 216
an intellectual and artificial fire 277
living fire 281
aethereal seminary 282

Chapter 8
"*Scire per causas*" Versus "*scire per signa*": George Berkeley and Scientific Explanation in *Siris*

Silvia Parigi

There are some topics, in Berkeley's philosophy, that have always puzzled his interpreters: first of all, the embarrassing concept of aether-spirit and the relations with contemporary science and with ancient hermetic wisdom as they result from his most puzzling, disordered and hermetic work.[1] To put *Siris* in order may seem a difficult enterprise, even if the subtitle of Berkeley's work is: *A Chain of Philosophical Reflections and Inquires Concerning the Virtues of Tar-Water, and divers other Subjects connected together and arising One from Another*. Therefore, Berkeley thought there was, or should be, an order in his last work: but the image he chose – the chain – refers more to Neoplatonic philosophy than to eighteenth-century science.[2] It reminds more Athanasius Kircher's universal magnetism[3] than the biological concept of *scala naturae*.[4] I will show that the structure of argumentation in *Siris* is hardly reducible to the classical inductive or deductive explanatory models, but nonetheless *Siris* is neither a "metaphoric, non-scientific, and often merely suggestive" work, nor "an epistemically humble text".[5] Berkeley just uses a

[1] David Berman defined it "Berkeley's most puzzling and allusive work": see his *Berkeley and Irish Philosophy* (London–New York: Continuum, 2005), 47.

[2] This is also the opinion of Timo Airaksinen: see his essay "The Path of Fire: The Meaning and Interpretation of Berkeley's *Siris*", in *New Interpretations of Berkeley's Thought*, ed. Stephen H. Daniel (New York: Humanity Books, 2008), 261–281.

[3] See the demanding work by the German Jesuit Athanasius Kircher: *Magnes sive de arte magnetica libri tres* (Romae: sumpt. Hermanni Scheus, ex typographia Ludovici Grignani, 1641).

[4] About which there is a wide bibliography, starting from the classical work by A.O. Lovejoy, *The Great Chain of Being* (Cambridge, MA: Harvard University Press, 1936).

[5] This is Timo Airaksinen's opinion: see "The Chain and the Animal. Idealism in Berkeley's *Siris*", in *Eriugena, Berkeley and the Idealist Tradition*, eds. Stephen Gersh and Dermot Moran (Notre Dame, Indiana: University of Notre Dame Press, 2006), 224–243 (quotations are from pages 237 and 228).

S. Parigi (✉)
Università di Cassino, Fac. Lettere e Filosofia, Via Zamosch 1, 03043 Cassino, Italy
e-mail: silpari@libero.it

S. Parigi (ed.), *George Berkeley: Religion and Science in the Age of Enlightenment*, International Archives of the History of Ideas/Archives internationales d'histoire des idées 201,
DOI 10.1007/978-90-481-9243-4_8,

different epistemological paradigm: not the most spread and historically winning causal paradigm (*scire per causas*), but the minor – though ancient – semantic paradigm (*scire per signa*).

What kind of order did Berkeley have in mind when he wrote *Siris*? What did he think of contemporary science and scientific explanation in general? What is the role of aether, or spirit, in Berkeley's epistemology? Can the history of these concepts cast any light on their meaning and connections? In this paper, I will try to answer these questions, reconstructing some links of the chain with the help of the history of ideas. In the first and second parts of my essay, I will analyze B's conception of scientific explanation in *Siris*, comparing two different models of science – "scire per causas" vs. "scire per signa" – both in their historical background and among Berkeley's contemporaries. In the third part, I will focus on the concept of aether, examining its role in *Siris* and showing its close relationship with the concept of spirit in Renaissance philosophy.

8.1 *"Scire per causas"*

To face contemporary science, in Berkeley's time, meant above all to confront Descartes and Newton, with their different epistemologies and models of the world. It is well-known that these philosophers elaborated two opposite cosmological systems, based on *plenum* and motion by contact (Descartes) or on vacuum and forces acting at a distance (Newton); it is not so stressed that they shared the same, classical, Platonic and Aristotelian concept of science as "scire per causas". That is to say, Descartes and Newton, the main scientific authorities in Berkeley's time, considered scientific explanation as a search for the true, efficient causes of natural phenomena. The only, though important, difference is that, in Descartes' opinion, efficient causes are of a mechanical kind: the motion, shape, size and figure of the smallest constituent parts of the physical bodies[6]; Newton didn't think so, but unsuccessfully tried to find the non-mechanical causes of the force of attraction. His long and unfruitful inquiry into the nature and cause of attraction highlights the importance that Newton gave to his research: a scientist cannot limit himself to state the laws according to which natural effects happen; if he speaks of "attraction", he has to say *what* it is, not only *how* it operates, because a formula can't work as a causal definition.

Among the few certainties that a careful reader of *Siris* is allowed to have, there are the following:

1. Berkeley strongly preferred Newton's physics. He energetically criticizes Descartes' system of the world:

[6] George Berkeley, *Siris*, in *The Works*, ed. A. A. Luce and T. E. Jessop (Edinburgh and London: T. Nelson and Sons, 1948–1957), vol. 6, § 243.

> Nothing could be more vain and imaginary than to suppose with Descartes that merely from a circular motion's being impressed by the supreme Agent on the particles of extended substance, the whole world, with all its several parts, appurtenances, and phenomena, might be produced by a necessary consequence from the laws of motion. (§ 232)

and enthusiastically praises Newton:

> Sir Isaac Newton, by his singular penetration, profound knowledge in geometry and mechanics, and great exacteness in experiments, hath cast a new light on natural science. The laws of attraction and repulsion were in many instances discovered, and first discovered, by him. He shewed their general extent, and therewith, as with a key, opened several deep secrets of nature, in the knowledge whereof he seems to have made a greater progress than all the sects of corpuscolarians together had done before him. Nevertheless, the principle of attraction itself is not to be explained by physical or corporeal causes. (§ 245)

Newton's system is set against the corpuscularian philosophy: in this opposition to mechanism lies the reason of Berkeley's preference.[7] Mechanical philosophers believe that matter exists, and that it is the ultimate cause of natural effects. On the contrary, on various occasions Newton explicitly refuses to attribute the force of gravity to matter. See, for example, the third letter to the apologist Richard Bentley, in 1692–1693, where Newton wrote:

> It is inconceivable, that inanimate brute Matter should, without the Mediation of something else, which is not material, operate upon, and affect other Matter without mutual Contact, as it must be, if Gravitation in the Sense of Epicurus, be essential and inherent in it. [...] Gravity must be caused by an Agent acting constantly according to certain Laws; but whether this Agent be material or immaterial, I have left to the Consideration of my Readers.[8]

Gabriel Moked spoke of a "radical change of mind" that Berkeley underwent when he wrote *Siris*, and that determined his "conversion" to "an hypothetico-deductive corpuscularianism".[9] On the contrary, I agree with Lisa Downing: Berkeley never did

---

[7] In opposition to most (if not all) Berkeleian scholars, Berkeley did not consider Newton's philosophy of nature as a mechanistic one. It is worth noting that many distinguished Newtonian scholars share Berkeley's opinion: see D.C. Kubrin, "Newton and the Cyclical Cosmos: Providence and the Mechanical Philosophy", *Journal of the History of Ideas* 28 (1967): 325–346; J.E. McGuire, "Force, Active Principles, and Newton's Invisible Realm", *Ambix* 15 (1968): 154–208; A. Thackray, *Atoms and Powers* (Cambridge, MA: Harvard University Press, 1970); Richard Westfall, *Force in Newton's Physics. The Science of Dynamics in the Seventeenth Century* (New York: Science History Publications, 1971) and "Newton and the Hermetic Tradition", in *Science, Medicine and Society in the Renaissance*, ed. Allen G. Debus (London: Heinemann, 1972), vol. 2, 183–198; B. J.T. Dobbs, *The Foundations of Newton's Alchemy* (Cambridge: Cambridge University Press, 1975); Ernan McMullin, *Newton on Matter and Activity* (Notre Dame: University of Notre Dame Press, 1978); Antonio Clericuzio, *Elements, Principles and Corpuscles. A Study of Atomism and Chemistry in the Seventeenth Century* (Dordrecht: Kluwer, 2000).

[8] Isaac Newton, *Papers & Letters on Natural Philosophy*, ed. I. Bernard Cohen (Cambridge, MA: Harvard University Press, 1958), 302–303.

[9] Gabriel Moked, "Two Central Issues in Bishop Berkeley's 'Corpuscularian Philosophy' in the *Siris*", *History of European Ideas*, 7 (1986): 633–641. See also his *Particles and Ideas. Bishop Berkeley's Corpuscularian Philosophy* (Oxford: Clarendon Press, 1988).

become a corpuscularian, nor adhere to a hypothetico-deductive model of science. Moreover, he did not search for or propose mechanical explanations of aether.[10] Nevertheless, he was not an "antiscience reactionary", but "a sincere and enthusiastic Newtonian",[11] who opposed Newton's "attractions and repulsions" to "those other mechanical principles of size, figure, and the like", and was convinced that "Nature seems better known and explained [...] by Sir Isaac Newton than Descartes" (§ 243); that "there appears a uniform working in things great and small, by attracting and repelling forces" (§ 234): "the different modes of cohesion, attraction, repulsion, and motion appear to be the source from whence the specific properties [of bodies] are derived, rather than different shapes or figures" (§ 162). I share Downing's assertion according to which: "Whereas in *De motu* Berkeley constructs a narrowly philosophical critique of the Newtonianism of the *Principia*, in *Siris* Berkeley produces a broadly philosophical meditation inspired by the Newtonianism of the Queries to the *Opticks*"[12] – with special regard to query n. 31, I would add: there are a lot of implicit citations from it in many paragraphs of *Siris*.

And now, let's come to the second main sentence that can be singled out concerning Berkeley's epistemology in *Siris*:

2. Both Newton's attractions and Descartes' corpuscles have the same epistemological status: in Berkeley's opinion, they are entirely destitute of causal power, therefore they are not to be assumed as causal principles, but only as "mathematical hypotheses, and not as anything really existing in nature" (§ 234).

> There is not any proof that an extended corporeal or mechanical cause doth really and properly act, even motion itself being in truth a passion. [...] We are not therefore seriously to suppose, with certain mechanic philosophers, that the minute particles of bodies have real forces or powers, by which they act on each other, to produce the various phenomena in nature. [...] The mechanical philosopher, as hath been already observed, inquires properly concerning the rules and modes of operation alone, and not concerning the cause; forasmuch as nothing mechanical is or really can be a cause. In conclusion, [...] if the explaining a phenomenon be to assign its proper efficient and final cause, it should seem the mechanical philosophers never explained anything.[13]

This is a powerful attack against Cartesian and corpuscularian philosophers in general; but the same criticism may be made of the Newtonians:

> Attraction cannot produce, and in that sense account for, the phenomena, being itself one of the phenomena produced and to be accounted for. [...] Attraction and repulsion should be considered only as tendencies or motions, that is, as mere effects, and their laws as laws

[10] The opposite opinion is maintained by Timo Airaksinen, in his paper on "Berkeley and Newton on Gravity in *Siris*", published in this volume; he speaks about different types of mechanisms in Berkeley's philosophy.

[11] Lisa Downing, "Berkeley's Natural Philosophy and Philosophy of Science", in *The Cambridge Companion to Berkeley*, ed. Kenneth P. Winkler (Cambridge–New York: Cambridge University Press, 2005), 253.

[12] Downing, "Berkeley's Natural Philosophy", 254.

[13] *Siris*, §§ 155, 235, 249, 231.

of motion. [...] The words attraction and repulsion may, in compliance with custom, be used where, accurately speaking, motion alone is meant. And in that sense it may be said that peculiar attractions or repulsions in the parts are attended with specific properties in the wholes. [...] The laws of attraction and repulsion are to be regarded as laws of motion; and these only as rules or methods observed in the production of natural effects, the efficient and final causes whereof are not of mechanical consideration.[14]

It should be noted that one possible source, in this case, is Newton himself:

I use that word [attraction] here to signify only in general any force by which bodies tend towards one another, and what are the laws and properties of the attraction, before we enquire the cause by which the attraction is performed. [...] These principles I consider, not as occult qualities, [...] but as general laws of nature, by which the things themselves are formed; their truth appearing to us by phenomena, though their causes be not yet discovered.[15]

In Berkeley's opinion, only phenomenal motions do exist *in rerum natura*, and they happen according to certain "rules and modes of operation" which, to speak properly, have the same meaning as "mathematical hypotheses". Therefore, the laws of attractions and repulsions, or of the impact of bodies, are mathematical hypotheses, useful to enable us to know what to expect: "Mechanical laws of nature or motion direct us how to act, and teach us what to expect" (§ 234).

Berkeley's conception of science in *Siris* is based on few paragraphs: in § 228 he describes the classical method of induction and deduction; in §§ 247, 252, 253 and 254 he states his position, which is not reducible to the previously mentioned methods. But let he himself speak:

It is one thing to arrive at general laws of nature from a contemplation of the phenomena, and another to frame an hypothesis, and from thence deduce phenomena. Those who supposed epicycles, and by them explained the motions and appearances of the planets, may not therefore be thought to have discovered principles true in fact and nature. And, albeit we may from the premises infer a conclusion, it will not follow that we can argue reciprocally, and from the conclusion infer the premises. [...] Though it be supposed the chief business of a natural philosopher to trace out causes from the effects, yet this is to be understood not of agents, but of principles, that is, of component parts, in one sense, or of laws or rules, in the other.[16]

A natural philosopher, therefore, may follow two ways: from the careful observation of phenomena, by a sort of Baconian induction, he may infer the rules of their production; otherwise, he may invent an hypothesis, and from this deduce phenomena, in a Cartesian way. Newton used (or should have used, according to his own epistemological principles) the first method: we should not forget his main slogan, expressed in the *Scholium generale* to his *Principia*: "Hypotheses non fingo", a motto that Berkeley should appreciate. In *query* 31 Newton wrote: "And although the arguing from experiments and observations by induction be no

[14] *Siris*, §§ 243, 246, 240, 231.

[15] Isaac Newton, *Optics*, *query* 31 (Chicago London Toronto: Encyclopaedia Britannica, 1952), 531, 542.

[16] *Siris*, §§ 228, 247.

demonstration of general conclusions, yet it is the best way of arguing which the nature of things admit of".[17] On the contrary, Descartes and his followers, "mechanic philosophers and geometricians" have been "misled by prejudice, and *have taken* mathematical hypotheses for real beings existing in bodies, so far as even *have made* it the very aim and end of their science to compute or measure such phantoms" (§ 250).

Anyway, sometimes Newton dangerously shifts close to Descartes' hypotheses: for instance, when he attributes a wide variety of phenomena – the attraction of iron towards the loadstone, straws towards amber, heavy bodies towards the earth, the irradiation of the sun, the repulsion between oil and water, mercury and iron, the double refraction of Icelandic crystal (§ 227), the chemical operations of cohesion, dissolution, coagulation, animal secretion, fermentation – to "the density and elasticity of aether", this "seems incomprehensible" (§§ 236–237).

> To explain cohesion by hamate atoms is accounted ignotum per ignotius. And is it not as much so to account for the gravity of bodies by the elasticity of aether? [...] Such a medium – distinct from light or fire – seemeth not to be made out of any proof, nor to be of any use in explaining the phenomena. [...] Attraction is performed by different laws, and cannot therefore in all cases be the effect of the elasticity of one uniform medium. The phenomena of electrical bodies, the laws and variations of magnetism, and, not to mention other kinds, even gravity, is not explained by elasticity, a phenomenon not less obscure than itself.[18]

Not diversely from Descartes' corpuscles, Newton's aether is an hypothesis, and hypotheses may be used only as assertions of laws or rules; they are to be taken as instruments to account for the ordered production of phenomena, whose efficient Cause is to be searched for in another realm: metaphysics or theology.

Moreover, if we have to choose an assumption for this aim, it doesn't seem useful to introduce a new principle, when we have an ancient one, ennobled by a long and glorious tradition of thought:

> It doth not seem necessary, from the phenomena, to suppose any medium more active and subtle than light or fire. Light being allowed to move at the rate of about ten millions of miles in a minute, what occasion is there to conceive another medium of still smaller and more moveable parts? Light or fire seems the same with aether. So the ancients understood, and so the Greek word implies. It pervades all things, is everywhere present. And this same subtle medium, according to its various quantities, motions, and determinations, sheweth itself in different effects or appearances, and is aether, light, or fire. [...] We are not therefore obliged to admit a new medium distinct from light, and of a finer and more exquisite substance, for the explication of phenomena which appear to be as well explained without it. [...] The phenomena of light, animal spirit, muscolar motion, fermentation, vegetation, and other natural operations, seem to require nothing more than the intellectual and artificial fire of Heraclitus, Hippocrates, the Stoics, and other ancients .... This animated heterogeneous fire should seem a more adequate cause, whereby to explain the phenomena of nature, than one uniform aethereal medium.[19]

---

[17] Newton, *Optics*, 543.

[18] *Siris*, §§ 227, 238, 243.

[19] *Siris*, §§ 226, 227, 229.

Neither induction nor deduction let the scientist find the true causes of natural effects; science is not a causal undertaking, it is rather an hermeneutic enterprise based on analogical processes of thought. Therefore, the links of the cosmic chain are not causal relations: analogy only permits to pass from one to another. Nature is not governed by "blind fate" or by "blind chance" (§ 273): causality and casualness equally fail to account for the grammar rules that the scientist has the unique and perhaps humble, but useful, task of discovering. Determinism and corpuscularianism may not be used as "influential metaphysics" for science; in the same way, deduction and induction do not work as explanatory models.

8.2 *"Scire per signa"*

The importance of Berkeley's doctrine of signs for an authentic comprehension of his philosophy has been generally admitted by scholars.[20] While acknowledging that "the doctrine of signs runs through, and to some extent unifies, almost everything Berkeley wrote", Henry Winkler denounces the lack, in Berkeley's philosophy, of a "fully general account of the nature of signs".[21] I suspect that the usual oblivion of the history of ideas may be responsible for this avowed deficiency. In Greek philosophy, "sign" is the opposite of "cause": it means a kind of inferential knowledge based on analogy and bearing to the formulation of conjectures.

There were two paradigms of scientific explanations deriving from ancient Greek philosophy and still used and familiar in Berkeley's time: *scire per causas* means to search for the efficient causes of phenomena, by means of hypotheses from which the effects may subsequently be deduced. This former model of science was first formulated by Plato in *Menon* (98a) and by Aristotle in the *First Analitics* (see, for instance, 33, 88b–89a) and *Second Analitics* (I, 13, 78b; I, 14, 79a; I, 31, 87b–88a; II, 19, 100b). The latter paradigm – *scire per signa* – was radically antithetic, and stemmed from the Stoic philosophy: a demonstration is the discovery of the consequent in reasonings of this sort: "If there is smoke, there is fire". We have an event, that stands for something else, and permit us to guess at its meaning, by conjectures based on analogical ties.

The main difference is that the relation between sign and thing signified is not necessary, but rests on the ground of repeated experience, and the inference is always a fallible process; on the contrary, cause and effect are linked by a kind of necessity that can't be founded on experience alone (as Hume was to show later on). Therefore, in Berkeley's opinion a scientist does not have the task of explaining phenomena: causal explanations belong exclusively to the metaphisician. The natural philosopher has the specific aim of comprehending the world around him,

---

[20] See for example, in this volume, Daniel's, Hight's, Kail's and Schwartz's papers.

[21] Kenneth P. Winkler, "Berkeley and the Doctrine of Signs", in *The Cambridge Companion to Berkeley*, 159–160.

by deciphering the characters that compose it, and that are not exclusively or mainly mathematical (as Galileo thought). He has to discover the rules of the grammar, in order to decode the language in which the book of nature is written, and make forecasts: therefore, his comprehension will only be analogical, hypothetical and inferential, not causal or certain.[22]

Berkeley had already stated this opinion in his *Treatise Concerning the Principles of Human Knowledge* (1710), where he wrote:

> The connection of ideas does not imply the relation of cause and effect, but only of a mark or sign with the thing signified. The fire which I see is not the cause of the pain I suffer upon my approaching it, but the mark that forewarns me of it. [...] Those men who frame general rules from the phenomena, and afterwards derive phenomena from those rules, seem to consider signs rather than causes.[23]

The same thesis was developed in *Siris* (§§ 252–254), where Berkeley expounded the *pars construens* in his doctrine of scientific method:

> There is a certain analogy, constancy, and uniformity in the phenomena or appearances of nature, which are a foundation for general rules: and these are a grammar for the understanding of nature, or that series of effects in the visible world whereby we are enabled to foresee what will come to pass in the natural course of things. [...] And in reality, he that foretells the motions of the planets, or the effects of medicines, or the result of chemical or mechanical experiments, may be said to do it by natural vaticination. (§ 252)

In this passage (from which I have omitted a significative reference to Plotinus' *Enneads*), it is to be noted that the paradigm of "scire per signa" based on analogy and conjecture (Berkeley uses the strongest expression of "natural vaticination") may be extended to every kind of science: not only the "Baconian" or inductive sciences (to use a famous label by Thomas Kuhn[24]), as medicine and chemistry, but also the mathematical or deductive ones, such as astronomy or mechanics. Therefore, this confirms what I said earlier, i.e. that the refusal of an explanatory, causal model of science puts induction and deduction equally out of the game. But let Berkeley go on:

> We know a thing when we understand it; and we understand it when we can interpret or tell what it signifies. Strictly, the sense knows nothing. We perceive indeed sounds by hearing, and characters by sight; but we are not therefore said to understand them. After the same manner, the phenomena of nature are alike visible to all; but all have not alike learned the connexion of natural things, or understand what they signify, or know how to vaticinate by them. [...] As the natural connexion of signs with the things signified is regular and constant, it forms a sort of rational discourse, and is therefore the immediate effect of an intelligent cause. [...] Therefore, the phenomena of nature, which strike on the

[22] As early as 1957 and 1961, two historians of ideas such as Max Jammer and Mary B. Hesse (who were not at all Berkeleian scholars) had fully realized and acknowledged this aspect of Berkeley's philosophy of science: see Max Jammer, *Concepts of Force* (Cambridge, MA: Harvard University Press, 1957), cap. 11 and Mary B. Hesse, *Forces and Fields* (Edinburgh and London: T. Nelson and Sons, 1961), cap. 7.

[23] George Berkeley, *A Treatise Concerning the Principles of Human Knowledge*, in *The Works*, vol. 1, §§ 65 and 108.

[24] Thomas Kuhn, "Mathematical vs. Experimental Traditions in the Development of Physical Science", in *The Essential Tension* (Chicago: The University of Chicago Press, 1997; first edition 1977).

> senses and are understood by the mind, form not only a magnificent spectacle, but also a most coherent, entertaining, and instructive Discourse; and to effect this, they are conducted, adjusted, and ranged by the greatest wisdom. This Language or Discourse is studied with different attention, and interpreted with different degrees of skill. But so far as men have studied and remarked its rules, and can interpret right, so far they may be said to be knowing in nature. A beast is like a man who hears a strange tongue but understands nothing. (§§ 253–254)

At this point, I would like to spend a word on the wrong inference, made by some Berkeleian interpreters, that the rejection of the causal paradigm of science means to renounce any predictability or possible explanation in natural sciences,[25] or is equivalent to a refusal of modern science *in toto*, with its method and objectivism. Our Bishop was not an adversary of Galileian and Newtonian mathematical physics, nor did he consider it only as a vain, artful and trifling grammatical exercise.[26] On the contrary, Berkeley accepted and admired contemporary science, though he (like Boyle) didn't privilege mathematical sciences: his conception of scientific explanation not catching the real causes of things, on the one hand is aimed at delimiting the scientific field from theology and metaphysics; on the other hand, such conception was not so odd among his contemporary philosophers, since it was shared, for instance, by John Locke and Robert Boyle.[27]

Let us give some examples from Berkeley's main philosophical source: in his *Essay Concerning Human Understanding* (1690), Locke wrote:

> I deny not, but a Man accustomed to rational and regular Experiments shall be able to see farther into the Nature of Bodies, and guess righter at their yet unknown Properties, than one, that is a Stranger to them: But yet, as I have said, this is but Judgement and Opinion, not Knowledge and Certainty. [...] Experiments and Historical Observations we may have, from which we may draw Advantages of Ease and Health, and thereby increase our stock of Conveniences for this life: but beyond this, I fear our Talents reach not, nor are our Faculties, as I guess, able to advance.[28]

In this passage, we find an asserted skepticism about our possibilities of authentic knowledge, that make Locke doubt the alleged causal certainty of natural philosophy, which "is not capable of being made a Science". Moreover, he stresses the practical, concrete advantages made possible by a kind of natural knowledge that can only be analogical:

> Concerning the manner of Operation in most parts of the Works of Nature: wherein though we see the sensible effects, yet their causes are unknown, and we perceive not the ways and

[25] This is Timo Airaksinen's opinion (see "Berkeley and Newton on Gravity in *Siris*", in this volume): "When the phenomena and their laws are really diverse and various so that their "mechanistic laws" are not available to us, only God can help. But, then we cannot predict."

[26] Denis Forest, "George Berkeley: langage visuel, communication universelle", *Revue philosophique de la France et de l'étranger*, 4 (1997): 429–446.

[27] On the history of the concept of sign in early modern philosophy of nature, see Massimo Luigi Bianchi, *Signatura rerum. Segni, magia e conoscenza da Paracelso a Leibniz* (Roma: Edizioni dell'Ateneo, 1987).

[28] John Locke, *An Essay Concerning Human Understanding*, ed. Peter H. Nidditch (Oxford: Clarendon Press, 1975), IV, 12, 10.

> manner how they are produced. We see animals are generated, nourished, and move; the Load-stone draws Iron; and the parts of a Candle successively melting, turn into flame, and give us both light and heat. These and the like Effects we see and know: but the causes that operate, and the manner they are produced in, we can only guess, and probably conjecture. [...] *Analogy* in these matters is the only help we have, and 'tis from that alone we draw all our grounds of Probability. [...] This sort of Probability, which is the best conduct of rational Experiments, and the rise of Hypothesis, has also its Use and Influence; and a wary Reasoning from Analogy leads us often into the discovery of Truths, and useful Productions, which would otherwise lie concealed.[29]

John Locke knew Robert Boyle, and Boyle's works, very well, and, without any doubt, Boyle had been a champion of this "wary reasoning" based on experience and analogy. Not diversely from Berkeley, he was also convinced that natural philosophy was only concerned with the second causes and their laws, and that it was aimed at promoting religious zeal and piety.[30]

Leibniz, too, agrees with Locke on this point: in the *Nouveaux essais sur l'entendement humain* (published posthumously in 1765, but composed in 1704–1705), his spokesman Theophilus says: "The art of discovering the causes of phenomena, or the authentic hypotheses, *is like the art of deciphering* [my italics], in which an ingenious conjecture often shortens the way".[31] Not casually, he soon cites Boyle as a practitioner of this genre of knowledge.

8.3 *Aether, or fire, or light i.e. spirit*[32]

Berkeley was not completely straightforward and coherent in his argumentation about his conception of science and the role of aether. The main objection might be: if aether is an hypothesis, like corpuscles, why should Berkeley prefer, and make use of, this hypothesis to account for the production of natural phenomena? There is a double answer to this question: first of all, as has already been said concerning attraction, aether is less involved with matter than mechanic corpuscles are.

> Fire seems the most elastic and expansive of all bodies. [...] This aether or pure invisible fire, the most subtle and elastic of all bodies, seems to pervade and expand itself throughout the whole universe. [...] Being always restless and in motion, it actuates and enlivens the whole visible mass. [...] So quick in its motions, so subtle and penetrating in its nature,

---

[29] Locke, *Essay*, IV, 16, 12.

[30] See, for example, Robert Boyle, *Some Considerations touching the Usefulness of Experimental Natural Philosophy* (1663), part I, in *The Works*, eds. Michael Hunter and Edward B. Davis, vol. 3, London: Pickering & Chatto, 1999.

[31] Gottfried Wilhelm Leibniz, *Nouveaux essais sur l'entendement humain*, IV, 12, 13, in *Die philosophischen Schriften* (Leibniz – Forschungsstelle der Universität Münster: Berlin, 1962), vol. 6.

[32] Contrary to Airaksinen ("The Path of Fire"), I think that these terms are assumed as synonyms by Berkeley, according to the Neoplatonic Renaissance tradition. I have explained the deep historical reasons for this synonymy more widely in "*Siris* and the Renaissance: Some Overlooked Berkeleian Sources", *Revue philosophique de la France et de l'étranger*, 1(2010): 151–162.

> so extensive in its effects, it seemeth no other than the vegetative soul or vital spirit of the world.[33]

That is to say, aether is less material than corpuscles, because it is "pure" and "invisible", never at rest, a sort of soul or spirit that animates the material "mass" of the world. Therefore, it should be preferred as an hypothetical principle, acting in nature as a secondary cause.

But this is not all: aether is to be preferred because it is an ancient principle, recently rediscovered by the greatest scientist in his time. It should be noted that the concept of spirit had been used as a theoretical alternative to mechanistic explanations also by some of Berkeley's remarkable contemporary philosophers, such as Henry More and Ralph Cudworth – besides Newton, of course.[34]

At this regard, Berkeley's humanistic erudition is decisive: in the paragraphs 166–168 of *Siris*, he sketches the history of this principle: Anaximenes, Heraclitus, Empedocles, the Pythagoreans, Hippocrates' *De diaeta*, Plato's *Timaeus* and the Platonists, pseudo-Aristotle's *De mundo* and the Peripatetics, the Stoics, Galen spoke about an original active principle of this kind.

> The Pythagoreans and Platonists had a notion of the true system of the world. They allowed of mechanical principles, but actuated by soul or mind. [...] They knew there was a subtle aether pervading the whole mass of corporeal beings, and which was itself moved and directed by a mind; and that physical causes were only instruments, or rather marks and signs. (§ 266)

In more recent times, the aethereal principle had been used again by Wilhelm Homberg, Herman Boerhaave, Bernard Nieuwentyt, Henry More, Ralph Cudworth and Isaac Newton.[35]

If *Siris* has always embarrassed Berkeley's scholars, it was also because Berkeley acknowledged two main authorities: the ancients and Newton, stating a connection between the doctrines of the Pythagoreans, Plato, the Stoics, Plotinus or Hermes Trismegistus and Newton's *queries*. It should be noted that Newton himself, as well as Berkeley, was deeply aware of the relationship between his philosophy of nature and that of the ancients. The Renaissance idea of a *prisca sapientia* was not foreign to him, especially as far as his own concept of light/spirit was concerned, as is testified in his 1675 letter to Oldenburg and some alchemical manuscripts, up to query 30 of *Opticks*.[36]

Berkeley powerfully claims that ancient wisdom deserves the attention of an authentic philosopher:

> The successful curiosity of the present age, in arts and experiments and new systems, is apt to elate men, and make them overlook the ancients. But [...] it must be owned that the

---

[33] *Siris*, §§ 149, 152.

[34] Daniel P. Walker, *Il concetto di spirito o anima in Henry More e Ralph Cudworth* (Napoli: Bibliopolis, 1986), 50.

[35] *Siris*, §§ 169, 189.

[36] See P.M. Rattansi, "Newton's Alchemical Studies", in *Science, Medicine and Society in the Renaissance*, vol. 2, 167–182.

ancients too were not ignorant of many things, as well in physics as metaphysics, which perhaps are more generally, though not first, known in these modern times. (§ 265)

It might still be objected, indeed, that also the corpuscular hypothesis is ancient and authoritative, since it arose from Democritus and Epicurus and was diversely reproposed and declined, in modern times, by Gassendi and Descartes. But this tradition of thought was notoriously impious and dangerous for the Christian doctrine and for theism of every kind. Moreover, it gave rise to the wrong, unacceptable Cartesian system of the world.

Blind fate [i.e. causal determinism in nature] and blind chance [i.e. casualness as the unique law of the impact of bodies] are at bottom much the same thing, and one no more intelligible than the other. Such is the mutual relation, connexion, motion, and sympathy of the parts of this world, that they seem as it were animated and held together by one Soul: and such is their harmony, order, and regular course, as sheweth the Soul to be governed and directed by a Mind. It was an opinion of the remote antiquity that the world was an animal. If we may trust the Hermaic writings, the Egyptians thought all things did partake of life. This opinion was also so general and current among the Greeks that Plutarch asserts all others held the world to be an animal, and governed by Providence, except Leucippus, Democritus, and Epicurus. (§ 273)

Therefore, causality may be double-faced, presenting itself both as spinozistic determinism and as epicurean atomism: in both versions, the causal paradigm of scientific explanation is to be rejected as dangerous for religion and inadequate for natural science. The ancient atomists were the only black sheep in the long and glorious chain of thought that considered the world an animated being,[37] or at least an ordered and harmonious *cosmos* produced and governed by a Mind acting through a pervading medium, which is aether or spirit, or fire, or light: "nor is this doctrine less philosophical than pious".[38] This line of thought runs through the whole history of philosophy, from the Greeks to Renaissance up to Newton's science. From this tradition, Berkeley utilizes the classical Neoplatonic topics of cosmic sympathy and antipathy, microcosm and macrocosm and of the spirit as a universal intermediary between soul and body (Marsilio Ficino is often quoted in *Siris*, and not by chance): in an original way (too much, for his critics, who have always failed to catch this relationship), he connects those concepts to Newton's aether.

Here are some passages concerning the correspondence between microcosm and macrocosm, based on the common presence of the spirit – be it the animal spirit as a medium of the soul acting on body and its limbs, or the cosmic spirit as an instrument of the soul of the world:

No eye could ever hitherto discern, and no sense perceive, the animal spirit in a human body, otherwise than from its effects. The same may be said of pure fire, or the spirit of

---

[37] T. Airaksinen (The Chain, 237) argues that Berkeley has a sort of "omnivorous" attitude towards ancient and modern philosophies and traditions of thought, using them to forge links in his cosmic and theoretical chain: he only excludes atheistic sources, such as Epicurus and Leibniz, "the archenemy of Newton".

[38] *Siris*, § 291.

> the universe, which is perceived only by means of some other bodies, on which it operates, or with which it is joined. [...] In the human body the mind orders and moves the limbs: but the animal spirit is supposed the immediate physical cause of their motion. So likewise in the mundane system, a mind presides: but the immediate, mechanical, or instrumental cause that moves or animates all its parts, is the pure elementary fire or spirit of the world. [...] that element, which, as it actuates the macrocosm, so it animates the microcosm.[39]

It might be useful to compare this Berkeleian passage to Newtonian definition of aether, considered the core of a unified theory of physical and physiological phenomena:

> a certain most subtle spirit which pervades and lies hid in all gross bodies; by the force and action of which spirit the particles of bodies attract one another at near distances, and cohere, if contiguous; and electric bodies operate to greater distances, as well repelling as attracting the neighboring corpuscles; and light is emitted, reflected, refracted, inflected, and heats bodies; and all sensation is excited, and the members of animal bodies move at the command of the will, namely, by the vibrations of this spirit, mutually propagated along the solid filaments of the nerves, from the outward organs of sense to the brain, and from the brain into the muscles.[40]

Moreover, Berkeley had ascribed the micro-macrocosmic "phenomena of light, animal spirit, muscolar motion, fermentation, vegetation" to the presence of an "intellectual fire".[41]

As for the cosmic sympathies and antipathies (or idiosyncrasies), they are audaciously linked to the Newtonian attractions and repulsions:

> Why may we not suppose certain idiosyncrasies, sympathies, oppositions, in the solids, or fluids, or animal spirit of a human body, with regard to the fine insensible parts of minerals or vegetables, impregnated by rays of light of different properties, not depending on the different size, figure, number, solidity, or weight of those particles, nor on the general laws of motion, nor on the density or elasticity of a medium, but merely and altogether on the good pleasure of the Creator, in the original formation of things? (§ 239)

Sympathies and antipathies are eminently analogical, not causal connections; the aether-spirit is the instrument through which they express themselves, and God is their ultimate, efficient and final cause. Contrary to what Newton thinks, there is no need to search for a physical cause of attraction; to account for natural phenomena, it is sufficient to adhere to the ancient theory of aetherial light or fire, considered merely as a second or instrumental cause of the various motions, whose laws are to be conjecturally, analogically stated by the scientist.

---

[39] *Siris*, §§ 159, 161, 166.

[40] This passage, drawn from the *Scholium generale* added to the second edition of *Philosophiae naturalis principia mathematica* (1713), is cited in Newton, *Papers & Letters*, 5. According to Walker (*Il concetto di spirito*, 18), Renaissance astrological and medical speculations about spirits were "an important, though neglected source ... of Newton's speculations on aether". In the same work, Walker reaffirms that "the whole tradition of material spirits could enlighten the interpretation of Newton's ideas on space and God and, strictly connected to those, his speculations about aether" (42).

[41] *Siris*, § 169.

Part III
Towards a Wider Historical Perspective

Chapter 9
Berkeley, Theology and Bible Scholarship

Daniele Bertini

Berkeley's immaterialism, in my opinion, has its main source in a reliable, personal, self-evident experience of God's presence in human existence. Contrary to other modern philosophers, who introduce the notion of a Supreme Being in order to solve substantial problems in their systems, Berkeley's attitude of mind seems near to the common religious feeling of a familiar and actual proximity of the divinity to man in everyday life.[1] Indeed there is no need to prove God's existence: "I am certain there is a God, though I do not perceive him and I have no intuition of him" (*Philosophical Commentaries*, 813).

This certainty is not the outcome of a geometrical demonstration: Berkeley explicitly denies the cogency of the ontological argument,[2] just as he claims the uselessness of all the traditional inferences from the world to God, although he admits

[1] My statement opposes the mainline of the traditional reception of Berkeley's philosophy. Scholars usually think that for immaterialism God is the typical *Deus ex machina* of the Cartesian age. My view refers to J.D. Mabbott, "The Place of God in Berkeley's Philosophy", *The Journal of Philosophical Studies*, VI, 1931; I.C. Tipton, *Berkeley. The Philosophy of Immaterialism*, Bristol: Thoemmes Press, 1994 (first edition, 1974), pp. 297 and following; M. Hooker, "Berkeley's Argument from Design", in C.M. Turbayne, ed., *Berkeley. Critical and Interpretive Essays*, Manchester: University Press, 1982.

[2] *Philosophical Commentaries*, 782. All quotations from Berkeley follow A.A. Luce and T.E. Jessop, eds., *The Works of George Berkeley, Bishop of Cloyne*, London, Edinburgh, Paris, Melbourne, Toronto and New York: Thomas Nelson and Sons, 1948–1957, IX volumes (hereafter quoted as *Works*).

D. Bertini (✉)
Department of Philosophy, University of Parma, Via Massimo D'Azeglio, 85/A 43100, Parma, Italy
e-mail: daniele.bertini@unipr.it

S. Parigi (ed.), *George Berkeley: Religion and Science in the Age of Enlightenment*, International Archives of the History of Ideas/Archives internationales d'histoire des idées 201, DOI 10.1007/978-90-481-9243-4_9,

their truth.[3] Experiencing God is a necessary consequence of a spirit's existence. For a spirit to be is to perceive.[4] Now, perceiving is to support the existence of sensible things, or, ontologically speaking, to be the *substratum* whose subsistence accounts for the actual *esse* of these.[5] Sensible things are therefore just determinations of substances, or spirits. But since any finite spirit *feels* that the totality of Nature is independent on its substantiality, so that he cannot be the ontological support of this totality, then Nature must be a modification of an infinite spirit, which is God. Things reveal in this way their spiritual nature, their springing from God's being. Things reveal the certainty of God's presence.

This revelation is the primal evidence of every existence, the very beginning of any spiritual life. Berkeley's dissatisfaction with the traditional ways to prove God's existence is, in my opinion, justified by his purpose to put God at the origin of personal experience. The Divinity is not to be found out in the frame of Nature, as something that can be inferred from the observation of the succession of causes and effects. No cosmological argument, such as arguments from beauty or usefulness, can offer men the right approach to God, since no reasoning can induce a spirit to have faith.[6] Before the truth of reason there is the absolute certainty of faith. Indeed

---

[3] *Three Dialogues between Hylas and Philonous*, II (*Works*, 2.212–213): "Divines and philosophers had proved beyond all controversy, from the beauty and usefulness of the several parts of the creation, that it was the workmanship of God. But that setting aside all help of astronomy and natural philosophy, all contemplation of the contrivance, order, and adjustment of things, an infinite mind should be necessarily inferred from the bare existence of the sensible world, is an advantage peculiar to them only who have made this easy reflexion: that the sensible world is that which we perceive by our several senses; and that nothing is perceived by the senses beside ideas; and that no idea or archetype of an idea can exist otherwise than in a mind. You may now, without any laborious search into the sciences, without any subtilty of reason, or tedious length of discourse, oppose and baffle the most strenuous advocate for atheism".

[4] *Philosophical Commentaries*, 429: "Existence is percipi or percipere…."

[5] *A Treatise concerning the Principles of Human Knowledge*, § 89: "*Thing* or *being* is the most general name of all, it comprehends under it two kinds entirely distinct and heterogeneous, and which have nothing common but the name, to wit, *spirits* and *ideas*. The former are *active, indivisible substances*: the latter are *inert, fleeting, dependent beings*, which subsist not by themselves, but are supported by, or exist in minds or spiritual substances".

[6] *Sermons*, IX (*Works*, 7.127–128): "The Christian religion was calculated for the Bulk of Mankind, and therefore cannot reasonably be supposed to consist in subtle and nice Notions. From the Time that Divinity was considered as a Science, and human Reason inthroned in the Sanctuary of God, the Hearts of its Professors seem to have been less under the Influence of Grace. From that Time have grown many unchristian Dissensions and Controversies, of men *knowing nothing, but doting about Questions and Strifes of words, whereof cometh Envy, Strife, Railings, evil Surmises, perverse Disputings of Men of corrupt Minds and destitute of Truth* (*1 Tim.* VI.4–5). Doubtless, the making Religion a notional Thing, hath been of infinite Disservice. And whereas its holy Mysteries are rather to be received with Humility of Faith, than defined and measured by the Accuracy of human Reason; all Attempts of this Kind, however, well intended, have visibly failed in the Event; and, instead of reconciling Infidels, have, by creating Disputes and Heats among the Professors of Christianity, given no small Advantage to its Enemies".

men do believe in the reality of their perceptions. The true world is what men experience.[7] There is not a *noumenon* beyond *appearance*. This appearance gives witness to inhere in God. Men experience being in relation among themselves by means of God's activity on Nature. God uses Nature to interrelate spirits. "Hence it is evident, that God is known as certainly and immediately as any other mind or spirit whatsoever, distinct from our selves .... He alone it is who *upholding all things by the Word of his Power* (*Heb*, I.3), maintains that intercourse between spirits, whereby they are able to perceive the existence of each other" (*A Treatise Concerning the Principles of Human Knowledge*, § 147). Berkeleian spirituality seems therefore near to the faith of the psalmist: "The Lord is my shepherd, I shall

[7] *Three Dialogues between Hylas and Philonous*, II (*Works*, 2.210–211): "Look! are not the fields covered with a delightful verdure? Is there not something in the woods and groves, in the rivers and clear springs that sooths, that delights, that transports the soul? At the prospect of the wide and deep ocean, or some huge mountain whose top is lost in the clouds, or of an old gloomy forest, are not our minds filled with a pleasing horror? Even in rocks and deserts, is there not an agreeable wildness? How sincere a pleasure is it to behold the natural beauties of the earth! To preserve and renew our relish for them, is not the veil of night alternately drawn over her face, and doth she not change her dress with the seasons? How aptly are the elements disposed? What variety and use in the meanest productions of Nature? What delicacy, what beauty, what contrivance in animal and vegetable bodies? How exquisitely are all things suited, as well to their particular ends, as to constitute apposite parts of the whole! And while they mutually aid and support, do they not also set off and illustrate each other? Raise now your thoughts from this ball of earth, to all those glorious luminaries that adorn the high arch of heaven. The motion and situation of the planets, are they not admirable for use and order? Were those (miscalled erratic) globes ever known to stray, in their repeated journeys through the pathless void? Do they not measure areas round the sun ever proportioned to the times? So fixed, so immutable are the laws by which the unseen Author of Nature actuates the universe. How vivid and radiant is the lustre of the fixed stars! How magnificent and rich that negligent profusion, with which they appear to be scattered throughout the whole azure vault! Yet if you take the telescope, it brings into your sight a new host of stars that escape the naked eye. Here they seem contiguous and minute, but to a nearer view immense orbs of light at various distances, far sunk in the abyss of space. Now you must call imagination to your aid. The feeble narrow sense cannot descry innumerable worlds revolving round the central fires; and in those worlds the energy of an all-perfect mind displayed in endless forms. But neither sense nor imagination are big enough to comprehend the boundless extent with all its glittering furniture. Though the labouring mind exert and strain each power to its utmost reach, there still stands out ungrasped a surplusage immeasurable. Yet all the vast bodies that compose this mighty frame, how distant and remote soever, are by some secret mechanism, some divine art and force linked in a mutual dependence and intercourse with each other, even with this earth, which was almost slipped from my thoughts, and lost in the crowd of worlds. Is not the whole system immense, beautiful, glorious beyond expression and beyond thought! What treatment then do those philosophers deserve, who would deprive these noble and delightful scenes of all reality? How should those principles be entertained, that lead us to think all the visible beauty of the creation a false imaginary glare?"

not be in want. – He makes me lie down in green pastures, – he leads me beside quiet waters – he restores my soul" (*Book of Psalms*, 23.1–3).[8]

Obviously it could be objected that no man has this immediate revelation of God. No one clearly and distinctly sees the Divinity shown by sensible things, when opening his eyes. A spirit could interpret its perceptions as it likes. Nonetheless, Berkeley has some (theological) reasons to hold his doctrine: the Irish philosopher is here following an argument proceeding from the Scripture.

In the *Gospel according to John*, the composer of the text writes: "The true light that gives light to every man was coming into the world. He was in the world, and though the world was made through him, the world did not recognize him" (*John*, 1.9–10). And furthermore: "This is the verdict: Light has come into the world, but men loved darkness instead of light because their deeds were evil" (*John*, 3.19). Berkeley refers explicitly to this topic of Johannine theology: "And yet this pure and clear light which enlightens every one, is it self invisible" (*A Treatise Concerning the Principles of Human Knowledge*, § 147). Men are to misunderstand the true nature of their perceptions because, even if they can perceive objects insofar as these are enlightened, they cannot perceive the light itself. Although they perceive the effects of God's activity, they do not immediately know the cause of all these sensations. They are free to choose whether or not to refer things to the Divinity. They are free to believe that things are signs actually pointing to God. All sensations suggest their having being constituted by God but cannot make him be seen by spirits as an object in the world, that is an idea. That is to say: since spirits have certainty that experience is a matter of contingency, Nature should be grounded by something overcoming the realm of experience (or not be grounded at all). Now, this putative ground of experience has the quality of transparency for the experiencing spirits (the same for the claim that contingency is all about the story), though being necessary to inquire on it. Thus, spirits have to query the meaning of their experiences. Berkeley lays down this claim, asserting that God is a spirit. God is actually near to every spirit, as Berkeley holds quoting Paul Acts 17.28. Nonetheless, this proximity is not knowledge like that concerning sensible things, since no spirit can have sensible ideas of other spirits.[9] Proximity of men and God is then better intelligible as a feeling, a mood, an acknowledgement by spirits to reduce things to God's will. Spirits need to grant by faith that what is shown in any perception is God.

---

[8] *Alciphron*, IV, § 7: "But if it shall appear plainly that God speaks to men by the intervention and use of arbitrary, outward, sensible signs, having no resemblance or necessary connexion with the things they stand for and suggest; if it shall appear that, by innumerable combinations of these signs, an endless variety of things is discovered and made known to us; and that we are thereby instructed or informed in their different natures; that we are taught and admonished what to shun, and what to pursue; and are directed how to regulate our motions, and how to act with respect to things distant from us, as well in time as place: will this content you?"; *Alciphron*, VI.27: "As for the Providence of God watching over the conduct of human agents, and dispensing blessings or chastisements, the immortality of the soul, a final judgment, and future state of rewards and punishments; how few, if any, of your free-thinkers have made it their endeavour to possess men's minds with a serious sense of those great points of natural religion".

[9] *A Treatise concerning the Principles of Human Knowledge*, § 148.

This faith totally crosses immaterialism. Berkeley speaks in some parts of his works using the language of epistemology and ontology, but his worries are mainly theological. And, in my opinion, it is not possible to understand his philosophy apart from this deep rootedness in the field of theology.

9.1 Berkeley's Notion of Theology: Science or Wisdom?

In any case, the notion of theology in Berkeley involves some difficulties, since textual evidences provide the reader with doctrines that vary in the different works. As to its definition Berkeley adopts a very traditional one: "But to treat of the good and great God, creator and preserver of all things, and to show how all things depend on supreme and true being, although it is the most excellent part of human knowledge, is, however, rather the province of first philosophy or metaphysics and theology than of natural philosophy which today is almost entirely confined to experiments and mechanics. And so natural philosophy either presupposes the knowledge of God or borrows it from some superior science" (*De Motu*, 34). Theology is at the top of the hierarchy of knowledge, regarding the source of everything and the relation between this source and everything. Its main concern is to know God as the real and true cause of everything that happens, and to comprehend the general frame of reality as immanent in God's activity.[10]

Apparently this is an acknowledgement of the soundness of the scholastic definition of theology as a positive science.[11] Is it then right to affirm that Berkeley's notion of theology comes from the scholastic one? Is the Irish thinker a follower of scholasticism with regard to this topic? Another passage would seem to confirm this reading. The entry 584 in the *Philosophical Commentaries* concerns the possibility to demonstrate in theology. Berkeley states that revealed theology accepts its principle by faith, but deduces all its consequences accordingly to a logical way of proceeding. "For tho' the Principles may be founded in Faith yet this hinders not but that legitimate Demonstrations might be built thereon. Provided still that we define the words we use & never go beyond our Ideas" (*Philosophical Commentaries*, 584). Aquinas claims something similar in his *Summa Theologiae*: theology is a science which goes along rationally proving its assertions by its principles (coming out from God's science).[12]

---

[10] *Siris*, § 285: "But those who, not content with sensible appearances, would penetrate into the real and true causes (the object of theology, metaphysics, or the philosophia prima), will rectify this error, and speak of the world as contained by the soul, and not the soul by the world".

[11] I use the term *Scholasticism* and its derivates to denote the general way philosophy was practised in the Universities during the eighteenth century. In this sense, it does not refer to Aquinas' philosophical system, but rather to the main concepts and definitions traditionally used by scholars in that age.

[12] *Summa Theologia*, I, *quaestio* 1, *art.* 2, *resp.*

Notwithstanding these formal resemblances, I don't think that the scholastic heritage is a good source for the immaterialist approach to the issue of theological knowledge. During the eighteenth century, the lasting scholastic tradition was still alive and in good health, being long rooted in the university institutions. For this reason its terminology, definitions, demonstrations and claims were technical devices for scholars. Now, even if this is the case, my claim is that all this does not prevent Berkeley from carrying out a kind of theology whose main tendency appears completely contrary to the scholastic one.

Indeed he explicitly denies, in the prosecution of the *Commentaries*' entry quoted before, that "to pretend to demonstrate or reason any thing about the Trinity is absurd. Here an implicit Faith becomes us" (*Philosophical Commentaries*, 584). There is an evident divergence between orthodox scholasticism and immaterialism: while Aquinas concludes his argument maintaining that theology is for its most part a speculative science, whose object is the knowledge of God's nature and properties,[13] Berkeley rather asserts that demonstrations grounded on Scripture are possible in moral and political subjects, but are of no concern in Metaphysics or First Philosophy.[14]

I would therefore be tempted to say that Berkeley accepts a scholastic framework for the treatment of the notion of theology as a common standpoint for scholars. But he then introduces radical innovations, first of all narrowing the use of theological demonstrations to the moral and political field of inquiry. This thesis appears plausible, even if some difficulties emerge in considering other textual passages concerning Catholic Theology. These passages show that Berkeley was not particularly interested in the possibility of building up a theological science. "The Scriptures and Fathers, I grant, are a much better help to know Christ and his Religion than the cold and dry writings of our modern Divines. Many who are conversant in such books I doubt have no more relish for the things of the Gospel, than those who spend their time in reading the immense and innumerable tomes of Scholastic Divinity with which the Church of Rome abounds. The dry polemical Theology was the growth of Rome, begun from Peter Lombard the Master of the Sentences, and grew and spread among the Monks and Friars under the Pope's eye" (*On the Roman controversy*, *Works*, 7.143). Apparently Berkeley states the uselessness of theology for an exact approach to faith. There is no need to practise a systematic kind of speculation concerning God, since all that the Christian requires in order to feed his faith is sufficiently set forward in *New Testament* literature. Eventually patristic commentaries supply useful integrations to the obscurities of the Holy Scripture. But surely, Berkeley claims, theological debates, controversies, squabbles seem far from enriching human experience and knowledge of God; the Bible suffices to this end.

---

[13] *Summa Theologia*, I, *quaestio* 1, *art.* 4, *resp.*

[14] *Philosophical Commentaries*, 585: "...Hence 'twere no very hard matter for those who hold Episcopacy or Monarchy to be establish'd *jure Divino*"; *On the Roman Controversy* (*Works*, 7.143): "The dry polemical Theology was the growth of Rome, begun from Peter Lombard the Master of the Sentences, and grew and spread among the Monks and Friars under the Pope's eye".

It should be noticed that the context of the quotation is deeply polemic: Berkeley is answering an apology of Catholic confession by Sir John James. His purpose is to attack a fundamental dogma of the Church of Rome: *extra ecclesiam nulla salus.* The object of his radical argument is the authority of the Pope and the usefulness of his guide to achieve salvation. It goes without saying that theological knowledge appears as the fruit of papal power: a scientific theology springs from the universities, institutions which appear to be the means to educate people in Catholicism. It could therefore be that Berkeley exaggerates his criticism of theology only for apologetic reasons. However, in my opinion, this is not the case.

The letter on Roman controversy offers a magisterial summary of Berkeley's thought. Indeed the doctrines laid down here had already been stated both in the *Treatise* and *Alciphron.*[15] Scholasticism is that paradigmatic kind of philosophy which corrupts faith with a misuse of reason. Like the ancient and modern freethinkers, the Schoolmen are so engaged with particularities that they cannot understand any kind of generality. Followers of the Scholastics are not able to see the moon pointed at by the finger, since they have too much interest in the finger itself. Berkeley explicitly describes this blindness of the knowledge coming out from the Schools: "There was indeed a time when Logic was considered as its own object: and that art of reasoning, instead of being transferred to things, turned altogether upon words and abstractions; which produced a sort of leprosy in all parts of knowledge, corrupting and converting them into hollow verbal disputations in a most impure dialect. But those times are past; and that, which had been cultivated as the principal learning for some ages, is now considered in another light" (*Alciphron,* V, § 24).

Therefore, taking into consideration all these statements together, it seems that Berkeley is completely unable to be consistent on this issue. On the one hand, he claims that theology is a science, whose method is demonstrative (but in maintaining this assumption he distinguishes himself from traditional Scholasticism); on the other hand, he denies the epistemological possibility to reason in Christian mysteries, since no knowledge can achieve the heights of faith.

---

[15] *A Treatise concerning the Principles of Human Knowledge, Intr.,* § 17: "It were an endless, as well as an useless thing, to trace the Schoolmen, those great masters of abstraction, through all the manifold inextricable labyrinths of error and dispute, which their doctrine of abstract natures and notions seems to have led them into. What bickerings and controversies, and what a learned dust have been raised about those matters, and what mighty advantage hath been from thence derived to mankind, are things at this day too clearly known to need being insisted on"; *Alciphron,* V, § 20: "No doubt all points in divinity are not of equal moment. Some may be too fine spun, and others have more stress laid on them than they deserve"; *Alciphron,* VII, § 9: "But all this may very justly be retorted on the minute philosophers themselves, who confound Scholasticism with Christianity, and impute to other men those perplexities, chimeras, and inconsistent ideas which are often the workmanship of their own brains, and proceed from their own wrong way of thinking. Who doth not see that such an ideal abstracted faith is never thought of by the bulk of Christians, husbandmen, for instance, artisans, or servants? Or what footsteps are there in the Holy Scripture to make us think that the wiredrawing of abstract ideas was a task enjoined either Jews or Christians? Is there any thing in the law or the prophets, the evangelists or apostles, that looks like it? Every one whose understanding is not perverted by science falsely so called may see the saving faith of Christians is quite of another kind, a vital operative principle, productive of charity and obedience".

It is worth noting that Berkeley approaches theology from a professional standpoint: first of all, he is a man of the church, who is well versed in theological issues. During his fellowship at Trinity College Dublin, he gave lessons in theology and exegesis of the *Old Testament.* His interest in the epistemology of theology, and in theology itself, was therefore strong and deep.

In confirmation of this claim, I refer to the apologetic argument for theology repeatedly advanced in *Alciphron.* Here his main intent is to differentiate between the truth of a discipline and the actual practice of it by men. While any knowledge demands just to be true, men in searching for truth are affected by different reasons. It would be a mistake to think that these reasons originate simply from a *thirst for knowledge.* Men often desire to become theologians, doctors or lawyers, in order to be respected by others or to obtain instrumental power. Men often quarrel, dispute or debate not to establish a truth, but to overcome their adversaries. Notwithstanding this, theorems and demonstrations of any knowledge remain true. No human behaviour could be an objection against truth. "But, after all, if men are puzzled, wrangle, talk nonsense, and quarrel about religion, so they do about law, physic, politics, and every thing else of moment. I ask whether, in these professions, or in any other where men have refined and abstracted, they do not run into disputes, chicane, nonsense, and contradictions, as well as in divinity? And yet this doth not hinder but there may be many excellent rules, and just notions, and useful truths, in all those professions. In all disputes human passions too often mix themselves in proportion as the subject is conceived to be more or less important. But we ought not to confound the cause of man with the cause of God, or make human follies an objection to divine truths" (*Alciphron,* V, § 19).[16] It does not seem then that Berkeley aims at criticizing the cogency of theological knowledge in general, but it could be the case that he simply maintains that scholastic theology is an unsatisfactory theological model in order to allow a true practice of theology.

My proposal to solve this difficulty is the following. I think the main problem in Scholasticism, for Berkeley, is the definition of theology as a positive science. Before being a kind of knowledge, philosophy and theology are indeed ways of life, concerning wisdom and truth. A philosopher should be a particular kind of man, like the theologian.[17] "Theology and philosophy gently unbind the ligaments that chain the soul down to the earth, and assist her flight towards the sovereign Good. There is an instinct or tendency of the mind upwards, which sheweth a natural endeavour to recover and raise ourselves from our present sensual and low condition into a state of light, order, and purity" (*Siris,* § 302). While the Schoolmen

[16] *Alciphron,* VII, § 9: "And it must be owned that the explication of mysteries in divinity, allowing the attempt as fruitless as the pursuit of the philosopher's stone in chemistry or the perpetual motion in mechanics, is no more than they chargeable on the profession itself, but only on the wrongheaded professors of it".

[17] *A Treatise concerning the Principles of Human Knowledge, Intr.,* § 1: "Philosophy being nothing else but the study of wisdom and truth, it may with reason be expected, that those who have spent most time and pains in it should enjoy a greater calm and serenity of mind, a greater clearness and evidence of knowledge, and be less disturbed with doubts and difficulties than other men".

claim to demonstrate God's presence as whatever objects of science, simply applying logics to theological matters, Berkeley believes that God is the meaning of any spiritual experience. Theology is not something that could be orderly displayed as a set of logically connected propositions. To take an example, I think Berkeley would have thought it rather foolish to write a handbook of theology. What matters in theological knowledge is just the personal attitude of the researcher: the tendency to bring any instance of existence back to God. Theology is the field of the experiences that men have of God. Theology is the experience of the divine.

The source of this experience, in the case of the Christian confession, is the Bible and the works by the Fathers. Since Scripture and religious literature do not concern the phenomenological world of ideas, but rather a simple revelation concerning an infinite Spirit – All is in God, comes from God and will return to God – which is experienced in the first person by holy writers, theology cannot be a science, but has to be an attempt to reduce things to God, to feel God as the focus of everything appearing in any personal experience. Thus theology is more similar to a kind of wisdom than to a scientific knowledge.

This wisdom concerns the fundamental movement towards God that any spirit has to make. Who is searching for God encounters wisdom: "Blessed is the man who listens to me, – watching daily at my door, – waiting at my doorway. – For whoever finds me finds life – and receives favours from the Lord" (*Proverbs*, 8.34–35).

9.2 David Berman's Reading of Berkeleian Notion of Theology

My reading contrasts with David Berman's representation of the structure of Berkeley's theology.[18] According to Berman, Berkeley holds that theology could be expressed in a twofold manner. The internal core of any religion is the belief in a wise and providential God. Metaphysics approaches this principle precisely and distinctly, laying down notions concerning the Supreme Being. Christian theology states mysteries beyond natural religion, whose justification could only be pragmatic by their emotive meaning. The believer cannot understand what the Trinity is, but can perfectly comprehend what Trinitarianism means from a pragmatic viewpoint.[19] The same experience of God could then be apprehended and spoken of in a cognitive or in an emotive way.

---

[18] D. Berman, "Cognitive theology and emotive mysteries in Berkeley's *Alciphron*", *Proceedings of the Royal Irish Academy*, 81 (7), 1981.

[19] *Alciphron*, VII, § 8: "Whence it seems to follow that a man may believe the doctrine of the Trinity, if he finds it revealed in Holy Scripture that the Father, the Son, and the Holy Ghost, are God, and that there is but one God? Although he doth not frame in his mind any abstract or distinct ideas of Trinity, substance, or personality; provided that this doctrine of a Creator, Redeemer, and Sanctifier makes proper impressions on his mind, producing therein love, hope, gratitude, and obedience, and thereby becomes a lively operative principle, influencing his life and actions, agreeably to that notion of saving faith which is required in a Christian".

In my opinion, this is not completely true, even if Berman rightly states that Berkeley distinguishes a twofold way of practising theological matters. In his reading, the object of rational theology – the existence of God – is the assumption that assures the possibility of revealed theology. The main problem with Berman's interpretation is the claim that this assumption could be an object of science. Natural theology would be the scientific knowledge that revealed theology would complete with an emotive treatment of Christian mysteries.

As I have already argued, I believe that Berkeley does not maintain that natural theology could be a science. Spirits know God's presence immediately. They feel as if they are in Him. "He has made everything beautiful in its time. He has also set eternity in the hearts of men; yet they cannot fathom what God has done from beginning to end. I know that there is nothing better for men than to be happy and do good while they live. That everyone may eat and drink, and find satisfaction in all his toil – this is the gift of God. I know that everything God does will endure forever; nothing can be added to it and nothing taken from it" (*Ecclesiastes*, 3.10–14). This religious attitude is common to all mankind. A natural tradition of faith could be traced back not only to heathens and Jews, but to all religious cultures too.[20]

According to my way of thinking, Berkeley claims that theology could not know anything more than this universal mind's mood regarding the relationship between appearance and truth. Naturally this is not a narrowing assumption: this kind of knowledge provides philosophers and theologians with an infinite field of inquiry. Indeed it requires a theological interpretation of personal experience, whose making needs the treatment of traditional theological doctrines.[21] But this kind of work is completely different from scientific research. That is why it would be nonsensical to reason about the Trinity, while it seems plausible to demonstrate in moral and political matters. Morals and politics concern life and action: men are actually furnished with ideas regarding the objects of their behaviour. Thus if a man believes a certain revelation, it would make sense to logically deduce all the pragmatic consequences of the revelation itself. As Berkeley claims, it suffices that the demonstration goes along accordingly to ideas whose ultimate certainty is proved by an actual perception.[22] In a certain way, it could be so granted that the realm of practice admits a scientific treatment. On the contrary, there are no perceptual handholds to reason into mysteries. They are revealed by God in Scripture. All theologians can explain them. But these explanations could not ever be anything else, other than determinate interpretations of their personal experiences of the divine.

---

[20] C. Bradatan, "Rhetoric of Faith and Patterns of Persuasion in Berkeley's *Alciphron*", *Heythrop Journal*, 47, 2006, pp. 552 and following.

[21] I think *Alciphron* is an interpretation of this kind, and the difference between early and mature works by Berkeley should be interpreted as the acknowledgement of the speculative poverty of a simply scientific or metaphysical explanation of theological matters. Here a cultural investigation is required.

[22] *Philosophical Commentaries*, 731a: "...real certainty is of sensible Ideas pro hic & nunc"; *Philosophical Commentaries*, 740: "We must with the Mob place certainty in the senses".

Therefore, I finally propose the following structure for Berkeley's notion of theology: every man has a natural experience of God. This is not knowledge but certainty grounded on experience, so that all natural or rational theology can know is simply that human existence is grounded on God's subsistence. But this is not the only source of theological experience, since God reveals himself directly to mankind through the Holy Scriptures. This revelation is not always clear, even if it is sufficiently understandable to achieve salvation. Thus any theological practice of reflection on the revelation can only be a kind of experience of the divine, whose subject is that of the theologian.

9.3 The Twofold Source of Theological Experience

A textual passage confirms that Berkeley explicitly claims a twofold source of theological experience in a systematic work in the field of theology. Coming back to the letter *On the Roman Controversy*, after having asserted the primacy of the Scripture and the Fathers as to the knowledge of Christian religion, Berkeley states that no man needs the Pope's guidance in theological issues, since a divine gift assists all men in correctly judging the truth of their beliefs. "There is an inward light... And a Christian soul wherein there is faith, humility, and obedience, will not fail to see the right way to salvation by that light which lightens the Gentiles and is a glory to Israel" (*On the Roman Controversy*, *Works*, 7.145).

This inward light is common to mankind. It is a natural ability with which God provides all spirits. Light proceeds from God to man. This event allows men to know. "God is the common father of lights; and all knowledge really such, whether natural or revealed, is derived from the same source of light and truth" (*Alciphron*, V, 9).

Having furnished men with light, God offers two kinds of revelation in the form of a twofold discourse addressed to any spirits. Accordingly, Berkeley affirms that the use of the inward light is offered to men together with an "interior as well as exterior *logos*".[23] In my opinion, the interior *logos* is to be interpreted as the spiritual experience of inhering in God, while the exterior is to be considered as the revelation spread by the Scriptures.

What is the theological meaning of these two different revelations? Do they have the same importance? Are they consistent with one another? My main claim is that Berkeley's theological system is constituted by the dialectic relation among the twofold sources of the divine revelation.

As to the first question, I would answer as follows: the first revelation is grounded on the metaphysical frame of human existence, while the second one is an historical event. The first is universal and concerns all men. The second is particular and affects just the followers of Christ's Gospel. The first is rooted into a divine tradition that all religious cultures have known. The second teaches the right

[23] *On the Roman Controversy*, *Works*, 7.145.

way to interpret the immanent revelation in any spiritual existence more clearly than any other theological insight. To summarize, it could be said that the first revelation is an ontological, or existential, experience of God, accounting for the cogency of natural theology; while the second is the appearance of the supernatural in human history, enabling us to modify the natural experience of God into a more enlightened experience of his attitude towards creatures.

At this point, a complete treatment of Berkeley's theology would require a detailed consideration of the doctrines advanced in his works, in the light of the twofold source of revelation. My purpose in the remaining part of my paper is to address topics raised by an insightful interpretation of Berkeley's theological thought, in order to further clarify the relation between the two kinds of divine revelation.

In a recent article, J.S. Spiegel argued that Berkeley's metaphysics is not only consistent with Scripture, but in some sense also recommended by some holy texts.[24] In the conclusion of his paper, he writes: "Berkeleian immaterialism enjoys at least as much and perhaps more explanatory power than matterism when approaching key biblical passages as the Genesis account of creation" (p. 231). The argument for this claim is that Berkeley's *divine language* theory would be particularly consistent with the "speech imagery" framing the cosmological text of the *Old* and *New Testament.* For this reason Spiegel concludes that immaterialism is perfectly compatible with orthodox Christian theology.[25]

In my opinion, the problem is that Berkeleian philosophy could also be consistent with some biblical passages; but even if this be the case, there are no warranties that immaterialism is theologically orthodox. In fact: firstly, Scripture does not provide readers with a unitary, uniform and homogeneous theology.[26] The Holy Bible collects texts written over about 1,000 years, attesting different experiences of faith, so that quoting some passages from the Bible is a very poor demonstration of orthodoxy. Eventually, a quotation may suggest a similarity of viewpoints or allude to a spiritual heritage, but it certainly cannot prove anything.

Secondly, Berkeley's immaterialist principle, whose consistency with the Bible Spiegel intends to prove, is previous to any revelation, being an achievement of God grounded on human knowledge. This foundation, coming from the world, is precisely what many theologians declare to be a complete refutation of the true Christian spirituality, since no rational, or philosophical approach to God could ever handle the primacy of the infinite on the finite. Tertullian among ancient Fathers, Luther among modern reformers and Moltmann among present-days

---

[24] J.S. Spiegel, "The Theological Orthodoxy of Berkeley's Immaterialism", *Faith and Philosophy,* 13 (2), 1996.

[25] While Spiegel claims that theological orthodoxy must be understood in terms of accordance with traditional ecumenical creeds, he grounds his argument for orthodoxy in the consideration of the Bible passages that could be consistent with immaterialism or that seem to require an immaterialist reading.

[26] G.Von Rad, *Theologie des Alten Testaments,* München: Chr.Kaiser Verlag, 1965, *passim*; E.S. Gerstenberg, *Theologien im Alten Testament. Pluralität uns Synkretismus alttestamentlichen Gottesglaubens,* Stuttgart: Verlag W. Kohlhammer, 2001, *passim.*

theologians, to give some examples, oppose the philosophical Athens to the eager and prayerful theological Jerusalem.[27] Philosophy and theology appear to be in opposition to large domains of Christian thought; and the doctrine of the twofold source of revelation, according to Berkeley's notion of theology, appears to be internal to this kind of opposition.

The problem of the orthodoxy of immaterialism should then be faced in the evaluation of the consistency among the two kinds of revelation for which Berkeley offers arguments. In this regard it should be argued as follows.

Some scholars, as well as myself, have expressed a view considering Berkeley representative of a particular Christian theological tradition, i.e. Christian Platonism.[28] The endeavour to reconcile Platonism with Christianity, in order to philosophically inquire into speculative difficulties provided by Scripture, is peculiar to this way of philosophizing. Berkeley refers directly to this tradition: "... several Fathers of the Church have thought fit to illustrate the Christian doctrine of the Holy Trinity by similitudes and expressions borrowed from the most eminent heathens, whom they conceived to have been no strangers to that mystery" (*Siris*, § 363). Many arguments could prove that Berkeley thinks of Platonists when he is claiming the agreement between *eminent heathens* and the Fathers. The most important argument is the massive use of the light analogy, a typical element that recurs especially in late Platonism, particularly in Christian Platonists such as Gregory of Nissa, Pseudo-Dionysius the Aeropagite or Maximus the Confessor.[29] For these authors Platonism and Christianity are two different ways of experiencing God. Is it orthodoxy? I don't know: it could possibly be the case that nobody can decide whether a doctrine is orthodox or heterodox. In any case, Berkeley seems to think that an unending theological tradition, coming directly from God, enables men to achieve the divine within or outside themselves. I am not so impressed by the claim that Platonists share the same opinions as Christians in theological matters, but I feel quite sure that Berkeley, like other Christian Platonists, was sincerely convinced

---

[27] Tertullian, *De praescriptione haereticorum*, 2; M.Luther, *Thesis debated in Heidelberg*, 19–22; J. Moltmann, *Der Gekrevzigte Gott*, München: Chr.Kaiser Verlag, 1973, *Passim*.

[28] N. Baladi, "Plotin et l'immaterialism de Berkeley. Temoignance de la Siris", Proceedings of the conference: *Plotino e il Neoplatonismo in Oriente e in Occidente*, Roma: Accademia Nazionale dei Lincei, 1974; P.S. Wenz, "Berkeley's Christian Neoplatonism", *Journal of the History of Ideas*, 37, 1976; K. Corrigan, "Berkeley and Plotinus on the Non-existence of Matter", *Hermathena*, 157, 1994; S. Daniel, "Berkeley's Christian Neoplatonism, Archetypes, and Divine Ideas", *Journal of the History of Philosophy*, 39, 2001; D. Bertini, *Sentire Dio. L'immaterialismo come via per un'interpretazione mistica dell'esperienza*, Assisi: Cittadella Editrice, 2007, *passim*; D. Bertini, "Mesta Panta Semeion. Plotinus, Leibniz, and Berkeley on Determinism", in P. Vassilopolou & S.R.L. Clark, *Late Antique Epistemology*, Houndsmills, Basingstoke, Hampshire: Palgrave Macmillan, 2009.

[29] W. Beierwaltes, "Die Metaphysik des Lichtes in der Philosophie Plotins", *Zeitschrift fürPhilosophische Forschung*, 15, 1961; W. Beierwaltes, *Denken des Einen*, Frankfurt am Mein: V. Klostermann, 1985, *passim*; W. Beierwaltes, *Platonismus im Christentum*, Frankfurt am Mein: V. Klostermann, 1998, *passim*; E. Moore, *Origen of Alexandria and St. Maximus the Confessor*, Boca Raton, Florida: Dissertation.com, 2005, *passim*; S. Lilla, *Dionigi l'Aeropagita e il Platonismo cristiano*, Brescia: Editrice Morcelliana, 2005, *passim*.

that it could be possible to show the consistency between Plato and Moses. Indeed the theological use of writings by Plato, Plotinus, Proclus, Iamblichus is justifiable for Berkeley by the twofold source of theological experience. The human experience accounts for a philosophical approach to God, while the revealed experience accounts for a theological one.

This doctrine is so deeply rooted in immaterialism that Berkeley sets it forth even in liturgical writings such as the Sermons: "It must indeed be owned that the Gentiles might by a due use of their reason, by thought and study, observing the beauty and order of the world, and the excellency and profitableness of vertue, have obtained some sense of a Providence and of Religion" (*Sermons*, IV, *Works*, 7.41). Furthermore, it is plain that both *Alciphron*' and *Siris*'s passages could provide evidence that Berkeley's purpose is to show the value of a God-oriented tradition, more ancient than Christianity itself. But it would be a mistake to deduce from this that the two kinds of revelation have the same importance.

Indeed, the human revelation may be understandable only to the higher spirits in the non-Christian societies. "But how few were they who made this use of their reason, or lived according to it! Perhaps here and there one amongst those who were called Philosophers: while the bulk of mankind, being diverted by the vain pursuits of riches and honours and sensual pleasures from cultivating their minds by knowledge and vertue, sunk into the grossest ignorance, idolatry and superstition" (*Sermons*, IV, *Works*, 7.41). Christ's coming into the world has enlightened more clearly the mind of all mankind. Christ was the perfect teacher: his mission was to enlarge the community of believers, formerly restricted to those well-read people who could correctly judge the source and meaning of their spiritual existence, as much as possible.

Now, since these teachings have been fixed in Scripture by holy writers, the study of Bible criticism appears fundamental to improve the exact knowledge of God's will. I will thus conclude my argument with a brief treatment of Berkeley's ideas concerning exegesis.

9.4 Berkeley's Defence of the Historical-Critical Method

Scholars of modern philosophy usually state assertively that scientific Bible criticism starts from the anonymous publication of Spinoza's *Tractatus theologico-politicus*. This appears to be a common belief, shared by all influential authors; so that it seems almost unnecessary to argue for it.[30]

In my opinion, however, this historiographical thesis is actually controversial. It is a bold claim asserted by the standpoint of a purely philosophical attitude to modern culture. On the one hand, this approach fails to acknowledge the existence

[30] R.H. Popkin, "Spinoza and Bible scholarship", D. Garrett, ed., *The Cambridge Companion to Spinoza*, Cambridge: Cambridge University Press, 1996, p. 383.

of a long history of literary studies exercised on the texts forming the canon of Scripture; on the other, it seems to disregard the fact that Spinoza's hermeneutical method is not only in large part really traditional, but also consistent with a genuinely Christian reading of the Bible.

At the start of the modern age, the study of the Scripture is an autonomous field of inquiry, extremely rich as far as reference sources are concerned. First of all, scholars had a great traditional commentary at their disposal: the *Latin Bible with ordinary Gloss*. The Gloss is a collective work coming out from the school of Anselm from Laon, who edited every single book in the Bible with introductory materials from S. Jerome and other commentators. These introductions are followed by the text with marginal or interlinear annotations by Anselm or some of his students.[31] Secondly, there were a large number of works concerning the philology of the Holy Scripture. In Christian culture, influential masters such as Stephen Langton reported in their *glosses* all the variations provided by different manuscripts, all the difficulties inferred from a comparison with Jewish rabbinical readings, all the possible divisions of the books into chapters and sections, in order to provide students with a common edition. Thirdly, scholars had the oversized *corpus* of commentaries, *disputationes*, *quaestiones*, inherited by the Fathers and the medieval theologians. Finally, after Luther's translation into German, versions of the Bible in modern languages began to appear.

The main problems of exegesis were the following: (a) whether the meaning of the texts is literary or allegorical; (b) whether the Scripture has to be the only canon to interpret Scripture, or a rational theology and philosophy are necessary premises of any readings; (c) whether the Scripture is clear or obscure; (d) whether the apparently obscured passages testify some difficulties in accepting the canon or, on the contrary, can have a theological justification.

According to scholars claiming that Spinoza's *Tractatus* is the groundwork of scientific Bible criticism, the Spinozistic revolution would consist of an assertion concerning the method to solve the second opposition laid down. P.F. Moreau writes in this regard: "It is necessary to treat the Bible as philology treats all other text, and, more generally, as the Science of Nature treats Nature. In this sense, interpreting Scripture on the basis of the only Scripture means that the interpreter chooses not to confuse his beliefs, or his personal choices, with the object which he is studying".[32] The main assumption justifying this method is that the meaning of the texts can be just literary and that Scripture presents obscurities that cannot be made sense of.

In my opinion, this position is well-attested in Christian and Jewish exegetical traditions. It suffices to read a medieval commentary on some biblical books to see how the alleged allegorical reading of the text is often really narrowed to single passages. When the text runs clear, readers usually tend to interpret it in a literal way. Furthermore, it is evident that no commentator tries to hide difficulties and

[31] B. Smalley, *The study of the Bible in Middle Ages*, London: Basil Blackwell & Mott, 1952, II, § 2.

[32] P.F. Moreau, *Spinoza. L'expérience et l'éternité*, Paris: PUF, 1994, p. 353.

inconsistent passages, usually approaching these obscurities with the tools of literary criticism. So, if this be the case, what differences do many scholars seem to see in Spinoza's work?

The problem concerns the reactions that the *Tractatus* provoked immediately after its publication. Its condemnation by the ecclesiastical authorities appeared to be a sign of the opposition between the despotical power of the Churches and the pretension to freedom of conscience affirmed by free-thinkers. In Spinoza's works, non-Christian philosophers could find a speculative account to justify their beliefs and to win over the prejudices of Christianity. Indeed what really impressed Christian theologians and philosophers was not the method chosen to read the Bible, but the assumptions concerning the reason why obscurities cannot be made sense of.

Berkeley's attitude towards Bible scholarship is paradigmatic of this kind of reaction. Berkeley does not feel any sympathy towards Spinoza, since he considers him to be the chief authority of free-thought. "Such, for instance, was Spinosa (*sic!*), the great leader of our modern infidels, in whom are to be found many schemes and notions much admired and followed of late years" (*Alciphron*, VII, § 26). Notwithstanding this, *Alciphron*, VI defends the view that literary criticism has to be the main method to interpret both profane and sacred books.[33] Here Berkeley openly acknowledges that many passages in the Scripture are corrupt, apocryphal, or mistakenly written out.[34] "That some few passages are cited by the writers of the New Testament out of the Old, and by the Fathers out of the New, which are not in so many words to be found in them, is no new discovery of minute philosophers, but was known and observed long before by Christian writers, who have made no scruple to grant that some things might have been inserted by careless or mistaken transcribers into the text from the margin, others left out, and others altered; whence so many various" (*Alciphron*, VI, § 7). In order to solve these difficulties, it is necessary to appeal to a better knowledge both of the languages in which the Scripture was written and of the social, cultural and religious context of the Jews and early Christians.[35] Therefore, Berkeley seems to accept the historico-critical method of reading the Bible without any difficulty. I do not think he would feel any contradiction

---

[33] *Alciphron*, VI, § 5: "Why then I would fain know whether it be equal and impartial in a free-thinker to measure the credibility of profane and sacred books by a different rule. Let us know upon what foot we Christians are to argue with minute philosophers; whether we may be allowed the benefit of common maxims in logic and criticism?"

[34] *Alciphron*, VI, § 5: "I know nothing truly valuable that hath not been counterfeited; therefore this argument is universal: but that which concludes against all things is to be admitted against none. There have been in all ages and in all great societies of men many capricious, vain, or wicked impostors, who for different ends have abused the world by spurious writings, and created work for critics both in profane and sacred learning. And it would seem as silly to reject the true writings of profane authors for the sake of the spurious, as it would seem unreasonable to suppose that among the heretics and several sects of Christians there should be none capable of the like imposture".

[35] *Alciphron*, VI, § 7: "And why will you not judge of Scripture by the same rule? Those sources of obscurity you mention are all common both to sacred and profane writings; and there is no doubt but an exacter knowledge in language and circumstances would in both cause difficulties to vanish like shades before the light of the sun".

in this, since he well knew that Christian hermeneutics provides many arguments in order to approach the Bible with the tools offered by literary criticism.

What is unacceptable in his eyes is the affirmation of the impossibility to clarify biblical obscurities. Spinoza holds that the divine inspiration depends on the prophet's ability to imagine. Even if God had spoken directly to the prophet, the prophet would have understood God's words according to his beliefs, state of mind and customs. A consequence of this premise is the fact that no rational or speculative principle regarding truth could have been diffused by Scripture. Berkeley denies that this explanation makes a good job. He is willing to admit that the prophet, being a man, could be the cause of misunderstanding in the receipt of God's speaking. But this does not affect the revelation, since this is not the main cause of biblical obscurities. Indeed Berkeley believes that the historical transmission of the holy text, as of any other, could account for its difficulties.

The Revelation is a fact concerning history: it comes in determinate ages, it is addressed to determinate men, is interpreted by determinate scholars, and is diffused by determinate churchmen. In this sense, revelation is exposed to all accidents and errors of other historical human events. It could be possible that a few important parts of the Bible, which are nonsensical nowadays, had a clear meaning in the past. It might also be possible that some unintellegible parts would become plain through the use of literary criticism. But what really matters, for Berkeley, is that the main ends of God could be understandable for men by the reading of the Bible.

These ends are confounded with the signs attesting that Scripture has a history. Readers must interpret the text in order to clarify its main theological insights. Readers must have an experience of the holy text; they have to show a relish for the Bible.

In this way, a parallel in the structure of natural and revealed theology is asserted. The foundation of both is the experiential nature of the existence of spirits. Spirits can exist only by experiencing something. But experiencing is to experience God. The natural theologian, or the philosopher, experiences God through his inward *logos*. The Christian theologian, as the simple believer, experiences God through his historical revelation in Scripture.

Chapter 10
The Distrustful Philosopher: Berkeley Between the Devils and the Deep Blue Sea of Faith

David Berman

10.1 Introduction[1]

Probably the main thesis I want to defend in this paper is one which I think few Berkeley scholars are likely to accept when first stated. My thesis is that Berkeley was a philosopher of little or no religious faith. But I hope that after reading further that some might be convinced and others at least open to the possibility that my thesis is true.

Now one thing that I do not intend to assert is that Berkeley was anti-religious or irreligious. I think he was religious, but that, like most people were at his age, place and time, he grew up assuming that the religion he was taught, namely Christianity, was true; and so, like most people, his acceptance of Christianity was originally based partly on faith or authority or habit or convention and partly on reason. But the difference in Berkeley's case was, I want to argue, that his religion became philosophical, that is, based entirely on reason. When did it become so? Although I do not pretend to date this exactly, I argue that it was there at least during those heroic years in Berkeley's life, *c.*1705–1707, when he was developing his immaterialist philosophy. So I hold that at least in that period his religion was based entirely on reason and not at all on faith, although I believe the evidence, which I shall present in Section 10.4, suggests that he was always a man of little or no religious faith. However, by focusing on this crucial period in Berkeley's life, I think my account takes on particular interest, since from the perspective of the history of philosophy

[1] I shall be using the A.A. Luce and T.E. Jessop edition of the *Works of George Berkeley*, nine volumes (London: Nelson, 1948–1957; *Works* for short) for all of Berkeley's writings, apart from his *Guardian* essays (see below note 15) and *Philosophical Commentaries*, where I will be using George Thomas's edition (Alliance, Ohio: 1976). I shall also be using the following abbreviations for Berkeley's works: PHK = *Principles*; DHP = *Three Dialogues*; PC = *Philosophical Commentaries*.

D. Berman (✉)
Philosophy Dept. Arts Building, University of Dublin, Trinity College, Dublin, Ireland
e-mail: dberman@tcd.ie

S. Parigi (ed.), *George Berkeley: Religion and Science in the Age of Enlightenment*, International Archives of the History of Ideas/Archives internationales d'histoire des idées 201,
DOI 10.1007/978-90-481-9243-4_10,

these are among the most important years of his life. Furthermore, I believe that my account can help to explain, in a psychologically satisfactory way, how Berkeley came to his revolutionary immaterialism. When I say 'satisfactory', what I have in mind is that where there is a powerful theory or belief – as there surely is in this case – it is more satisfactory to have a powerful cause for that belief.

Now one thing that I like about my thesis – and I hope others will also like – is that it goes against the condescending and common picture of Berkeley, that he was religious because he had to be – in short, because he was a clergyman and then bishop; and, more specifically, that it was because of his vocation and/or prejudice that he was strong minded against matter but weak minded with regard to spirit. In my view, on the contrary, Berkeley was through and through a strong believer in reason, that he was religious because he was a true philosopher who saw the truth of religion, and that he was not a philosopher because he was religious. So as I see it, he could even have been an atheist, if reason went that way. Of course, given what we know of him and the time and place he lived, I'm sure that such an option would not have been easy for him. And ultimately it might even have driven him crazy; but still, I think, he could and would have accepted it, because his meta-religion, so to speak, was reason and truth. So I believe that Berkeley was in the tradition, described and criticized by Nietzsche, according to which 'nothing is needed more than truth', a tradition which Nietzsche identifies with Plato.[2]

To keep things clear and simple, I shall also now say a word about what I understand by Berkeley's principal religious belief; which I take to be a belief in the God of theism, the God with all the usual intellectual and moral perfections, plus concern for the well being of man. This last attribute, which used to be called God's providence, is one which Berkeley makes much of, especially in *Alciphron* (1732) and the *Theory of Vision Vindicated* (1733). In this important respect, Berkeley's God goes beyond theism or deism and moves closer to the God of Christianity.

From what then do I derive my main thesis? Clearly if I am to have any chance of convincing scholars, I need to bring in some elements which they can accept. And I have two such main elements, the first of which is (1) that Berkeley opposed scepticism. Although few or no commentators would disagree with me in general about this, I press this much much harder than most; for I think that for Berkeley it was probably the most horrible and serious enemy, the deepest root of all evils – even worse than irreligion.[3]

The other main element from which I hope to derive my main thesis is (2) that eventually he found a way of dealing with scepticism, namely by his esse is percipi principle which issued in immaterialism. Here again, I doubt if any scholar would dispute that, but as I hope to show, it was much more than that for Berkeley, since

[2] See F. Nietzsche, *The Gay Science* (New York: Vintage, 1974), trans. and ed. by W. Kaufmann; see number 344; for Berkeley's own personal commitment to the truth, see the last section of *Siris* (*Works*, vol. 5, p. 164).

[3] See the end of *Alciphron*, where Berkeley traces the source of irreligious freethinking to scepticism; see *Works*, vol. 3, 316–329. In short, irreligion is bad but scepticism is worse, because it is scepticism that produces irreligion and not vice versa.

it was his way of avoiding the personal awfulness of scepticism or perhaps curing himself of something like a dark night of the soul, which he had to endure for a time.

10.2 The Unhappiness of Scepticism

So now I need to explain my first element. I could quote a lot of evidence about Berkeley's hostile attitude to scepticism, but there are three or four sources that I think are crucial. The first is Section 1 of the Introduction to the PHK, where Berkeley sets out in general terms his view of scepticism, describing what happens to philosophers when they depart from the ways of ordinary folk. And by philosophers, it is clear that he means to include what we would now call scientists. More specifically he also seems to be referring to those thinkers who accepted the New Science – of Galileo, Descartes, Newton, Boyle, Locke and others – not those who still adhered to the old Scholastic or Aristotelian way of thinking. The main way they differ from ordinary folk is that ordinary folk are calm and serene, whereas the philosophers and scientists have lost their natural serenity and calm. So ordinary folk are not 'disturbed with doubts and difficulties'. They walk the 'high-road of plain commonsense' and are 'for the most part easy and undisturbed'. Why? Because they believe their senses and are not in danger of becoming sceptics, which is different with philosophers. And here, interestingly, Berkeley switches to the first person plural: 'But no sooner do we [he writes] depart from sense and instinct to follow a superior principle', namely reason, then 'a thousand scruples spring up in our minds', which leads us into 'uncouth paradoxes' and 'through many intricate mazes, [till] we find ourselves just where we were, or, which is worse, [we] sit down in forlorn scepticism.'[4]

Now the main thing I want to call attention to is that what Berkeley is saying here is not that the scepticism of the philosophers is mistaken but that it leads to mental disturbance and probably a forlorn state of mind. This also comes out in the term 'scruples' which Berkeley uses in Section 1 and also in the first paragraph of the DHP, which is the counterpart of Section 1, where he describes those 'addicted to speculative studies', which 'perplex the plainest things, that distrust of the senses, those doubts and scruples....' Now in Berkeley's time the term scruples carried a more negative connotation or emotive meaning, indicating doubtfulness more than carefulness. Thus Shaftesbury uses scrupulist as a synonym for sceptic. But the term scuples could carry an even more negative meaning, as something

[4] When Berkeley says 'just where we were', I take it that he means back in the initial commonsense or ordinary condition that he characterizes as calm and serene. I think this reading is supported by, for example, the very last paragraph of the DHP, but especially paragraph seven of the Preface to the DHP, where Berkeley speaks of taking his readers on a 'curcuit through so many refined and unvulgar notions, [in which] they should at last come to think like other men: yet [he says] this return to to the simple dictates of nature, after having wandered through the wild mazes of philosophy, is not unpleasant', since a man can 'reflect with pleasure on the many difficulties and perplexities he has passed through...' (*Works*, vol. 2, 168.) Also see *Works*, vol. 9, p. 153 for a note to the same effect.

close to, if not virtually identical with, the anxiety disorder we now call obsessional compulsive disorder or OCD. Thus Jeremy Taylor says that

> A Scruple is a great trouble of mind proceeding from a little motive. That it is a great trouble is a daily experiment.... Some persons dare not eat for fear of gluttony, they fear they shall sleep too much, and that keeps them waking.[5]

And Taylor gives a case study to illustrate how uncertainty could become neurotic, even psychotic. As hysteria seems to have been the prime mental illness of the late nineteenth century, so scruples or religious OCD seems to have been the prime mental illness of the middle seventeenth century. Indeed, it was so common that in Oxford there were even what were called 'Scruple shops' for treating those suffering from it, shops that we would now call clinics. And though scruples were primarily religious in Taylor's day, he also recognizes how they can be bodily or psychological or concerned with virtually anything – 'from anything that may abuse the reason into irresolution and restlessness.' (p. 163). What I am suggesting is that for Berkeley scruples had moved on somewhat from being what Taylor calls 'a religious melancholy' and took an epistemic manifestation, suitable for the Age of Reason.

I now want to consider my second source which is from the PHK, Sections 86–88, where Berkeley fills in the general account in Section 1 of the Introduction by explaining in detail what specific reasoning leads to forlorn scepticism. This is the familiar account, according to which the philosophers and scientists have erected a twofold or representationalist order of existence, with the real things lying beyond the veil of perception, the effect of which is to make it impossible for us to know what really exists, or even to know that material things do exist. This is what I shall sometimes call the gap, by which I mean the gap representationalism or dualism creates between what we experience and what exists, which Berkeley also calls attention to in the first two paragraphs of the Preface to the DHP. And so it is the gap, as Berkeley says, that plunges us into a state of 'doubtfulness which so bewilders and confounds the mind' (Section 88). So here again, as in Section 1, he is emphasizing the unsettled, unhappy condition of the sceptic – so different from the calm and serene state of mind of commonsense people as depicted in Section 1.

This also comes out perhaps most clearly in my final source, that at the beginning of DHP, dialogue 3, where Hylas, now a thorough sceptic, professes to be ignorant of the true nature of all real things, even to the point of questioning if any real things exist, although he does not deny that he experiences the appearances. Philonous reaction to Hylas's sceptical confession is first to observe how 'wild and extravagant' Hylas's notions are, then, in line with PHK 86–88, to emphasize the source of Hylas's scepticism, that it comes from 'distinguishing between the reality and sensible appearance of things'. It is to this [says Philonous] that you are indebted for being ignorant of what everybody else knows perfectly well.'

> And so [concludes Philonous] you are plunged into the deepest and most deplorable scepticism that ever man was. Tell me, Hylas, is it not as I say?

To which Hylas replies: 'I agree with you.'

---

[5] In his *Ductor Dubitantium, or Rule of Conscience* (London: Royston, 1660); quoted in I. Macalpine and R. Hunter, *Three Hundred Years of Psychiatry* (London: Oxford University Press, 1963) 163–165.

To be 'plunged into the deepest and most deplorable scepticism that ever man was' is bad. But fortunately by the very end of dialogue 3 Philonous is able to cure Hylas of this deplorable scepticism by the *esse* is *percipi* principle and Immaterialism. And now that he is converted Hylas is not only able to accept that he was a sceptic, but he can also accept something else that he had not clearly recognized. Here I think Berkeley is making an acute and subtle psychological observation. Hylas puts it this way in his last speech:

> I have been a long time distrusting my senses; methought I saw things by a dim light, and through false glasses. Now the glasses are removed, and a new light breaks in upon my understanding. I am clearly convinced that I see things in their native forms; *and am no longer in pain about their unknown natures or absolute existence.* (*Works*, vol. 2, p. 262, my emphasis)

So now that his skepticism and the pain arising from it is over, Hylas is able to realize just how painful it was. And this is important, I believe, because it mirrored Berkeley's own experience, of the pain of scepticism. In short, Berkeley used his own experience to describe that of his creation, Hylas – just as novelists are known to use their own experience to construct that of their fictional characters. What we also have in Hylas is the epistemic equivalent of the sinner who, realizing the error of his ways, not only becomes good, but acutely aware of how unhappy he actually was in his previous immoral life.

However, the primary thing I am trying to draw attention to, which I think has been largely passed over in accounts of Berkeley, is the emphasis that he gives to the emotional state of mind of the sceptic: its forlorn, painful, disturbed, scruple-inducing, bewildering, deplorable character, which arises from the bifurcation of appearance and reality, i.e. the gap. Of course, as a philosopher, Berkeley's main concern is to show that the gap is unwarranted, unnecessary and mistaken; but he also wants his readers to appreciate how personally disastrous it is, how it leads to unhappiness. So he is like a physician who is explaining to his patient how smoking leads to illness, but also how wretched such illness is.

10.3 Who Were the Unhappy Sceptics?

Now at this point, I should like to pose probably the most important question of this essay. Although Berkeley describes in some detail the sceptics's theory and unhappy state of mind arising from it, he never tells us *who exactly these sceptics were or are. So who are they?* If some readers believe that there are plenty of candidates to choose from, then I shall be particularly happy, since that would convince me that this essay was not written in vain. But I believe that there are no such sceptics to be found.

10.3.1 Sextus Empiricus

Probably the best way for me to prove this and answer my key question is to go methodically through the most likely or prominent pre-1710 candidates. And that, as far as feasible within the scope of this essay, is what I now propose to do.

And what better candidate to begin with than probably the most straightforward sceptic, namely Sextus Empiricus, our main source for classic Pyrrhonic Scepticism? Could it be Sextus? Well, Sextus certainly accepts the distinction between reality and appearance and he is clear that so far he has not been able to go beyond appearances to understanding the real nature of things. For unlike the dogmatists, who he is opposing, Sextus accepts that so far he does not know any real things. So in those crucial respects he is exactly like Hylas at the beginning of dialogue 3.

Yet for all that, Berkeley description cannot be of Sextus and the Pyrrhonian sceptics, because according to them what especially distinguishes their position is that, as against that of their opponents, it brings about calm and serenity. That, Sextus holds, is the very *raison d'être* and main recommendation of their approach.[6] And the way the Pyrrhonian sceptic achieves this calm is by showing that any dogmatic claim about the real nature of X can be balanced by an equally compelling antithetical claim. This is their famous method of equipolence. The balancing of opposing dogmatic positions produces the desired calm. So if Berkeley had the Pyrrhonian sceptic in mind, then he was badly, indeed outlandishly wrong.[7]

In order to cut to the chase, let us come quickly to modern times. And in doing so, it might seem that we are also getting to or certainly closer to our desideratum, since the gap was the dominant epistemic theory of the new science, as can be seen in the position of the advanced thinkers of the time. (Indeed it was this fact that prompted Bayle to say that the new science vindicated the Pyrrhonian sceptics and showed that they had been right all along.[8]) Of course, clearly I cannot in this paper look at every significant advanced thinker of the seventeenth and early eighteenth century, so I shall restrict myself to the following: Descartes, Malebranche, Locke, Hobbes, Spinoza, and Bayle – all major thinker, who we know Berkeley had studied.

Now equally or even more than the Pyrrhonian sceptics these thinkers accepted the gap between our sensory experiences and material reality, which was perhaps the most important way that their position differed from their opponents, the Aristotelians and Scholastics, who accepted a realist theory of perception and epistemology.

This is the position perhaps first clearly articulated by Descartes, according to which there are two substances in the world, the mental which directly experiences ideas and sensory states and the material or extended, which we know indirectly by means of these states and reason. Here then in Descartes we have the basic problem

[6] All references to Sextus Empiricus are to *Selections from the Major Writings of Sextus Empiricus* (Indianapolis: Hackett, 1985), ed. by Philip P. Hallie; see 35–42.

[7] Berkeley's representation of the sceptic might seem to be closer to the Academic Sceptics, according to whom we can't know the real nature of anything. They differ from the Pyrrhonian sceptic, who says that he doesn't at present know any such thing. But here again there is no evidence that I know of that the Academic sceptic suffered from his scepticism or this invincible ignorance.

[8] For all references to and quotations from Bayle, see Richard Popkin's edition of the *Historical and Critical Dictionary: Selections* (Indianapolis: Hackett, 1991); see 196–197.

which Berkeley identifies as the root of scepticism. And although we need to look at the particular ways it manifests itself in subsequent philosophers, nearly all of them accepted it in different ways, which was largely forced on them by the new science, according to which the material world is quantitative, consisting of what became known as the primary qualities, whereas what we experience by sense, the secondary qualities or ideas of them, is qualitative and subjective. But then how can we move from what we directly experience to what really exists but which we know only indirectly? This was the problem of the gap. And the way Descartes and nearly all philosophers solved it was by invoking God, to the effect that since we know He is perfect we can be sure that He has organized the world so that our sensory states or ideas or representations do represent the material things.

10.3.2 Hobbes and Spinoza

However, there were two main philosophical exceptions to the prevailing dualism, namely Hobbes and Spinoza. They accepted the new science but found ways of integrating it into their non-dualistic metaphysics. In Hobbes's case, this was to hold that consciousness or what we immediately experience is in fact physical, basically motion. So, in short, materialism can eliminate the gap between the so-called mental and the physical realms, by eliminating the mental. Spinoza's position is more complicated, but before saying something about him, I want to describe the overwhelming dualistic consensus of the time. But here I need to remind the reader about what, in particular we are looking for. There is no problem finding plenty of 17th or early 18th philosophers or scientists who accepted the dualistic or representationalist problem, which Berkeley regarded as the root or recipe for scepticism. What we are looking for are those who experienced it in the way described by Berkeley, i.e., as fostering or issuing in a forlorn, disturbed, scruple-inducing, bewildering, deplorable state of mind.

10.3.3 Descartes

Let us begin then with Descartes. That Descartes's philosophy cuts us off from the material world in the way Berkeley describes should be clear from *Meditation* 2, where by his wax example, Descartes tries to show us that while we believe that we experience the real material wax by sensory perception, we are mistaken. What we experience by our senses is subjective and mutable. But there is a real non-mutable wax, the extended wax, and it is known by our intellect. And yet it is clear from *Meditation* 1, that there might be no real wax. For it might be that what exists are only our subjective experiences, as in a dream, or as brought about by the evil daemon. Now that possibility would be a most disturbing one, a nightmare possibility. But it is not what Descartes believes to be the case. For he believes that, beginning with cogito, he is able to prove that there is a perfect being, who is the cause of this world, and that this perfect being, God, would not deceive us about the existence and nature of the material world. This is the upshot of the *Meditations* and is given more formal and sober statement in Descartes's *Principles* II.1, where he begins by stating that 'Everyone is quite convinced of the existence of material things.' But he notes that 'we [have] cast doubt on this belief and counted it as one of the preconceived opinions of our childhood.'[9] So it is necessary, now that he and hopefully we have moved from the pre-critical position of child to the rationalist position based on the cogito, to see if there is a good argument or reason for the belief in material things.

Here he begins with the proposition that 'all our sensations undoubtedly come to us from something that is distinct from the mind. For it is not in our power to make ourselves have one sensation rather than another....' So there is something distinct from us or our minds which causes these involuntary sensations. Now this thing, distinct from the mind, he takes to be extended matter. And Descartes says in *Meditation* 6 that we human beings have a 'strong propensity' to believe that it is corporeal substance that produces our sensations (possibly being the first to identify this strong propensity). But he admits that it could be God that is causing these sensations and also causing our idea of extended matter. Yet in that case, he says, 'there would be no way of avoiding the conclusion that He should be regarded as a deceiver'. But 'this is quite inconsistent with the nature of God'. Therefore we can be sure that there are material things.

So for Descartes there is no reason to be disturbed or in the forlorn or deplorable state of mind, as described by Berkeley. We can feel serene, resting in the security of the most perfect being. But here a question arises about this ground of our serenity. If God, as a most perfect being, would not deceive us, then does it follow from this that God cannot deceive us, and hence that the existence of matter necessarily follows from his perfect nature? If so, then God is not free – which is the way that Spinoza developed his philosophy, drawing out the conclusion from the idea of God as the most perfect being. Spinoza, in other words, takes Descartes in a pantheistic direction. God becomes substance or nature, which expresses itself necessarily through thought and extension (and all possible attributes) and is known (to humans) in this dual way. But it is unlikely that Descartes himself would have accepted this Spinozistic development, because (1) it undermines his dualism and (2) his theistic conception of God. For unlike Spinoza, Descartes presumably believed that God is free. But if Descartes was not prepared to go the Spinozistic way, and God is free, then He might deceive us. But if He might deceive us, then should Descartes be so calm? Indeed, if He might deceive us, then it could be

[9] Quotations from Descartes are from *The Philosophical Writing of Descartes* (Cambridge: Cambridge University Press, 1987), trans. and ed. by J. Cottingham, R. Stoothoff, D. Murdoch; see vol. 1, p. 223. In his 16th Objection to Descartes, Hobbes points out that the imperfection of deception does not consist in the falsity of what is said but in the harm done by the deceiver; hence we do not blame a doctor for deceiving patients for the sake of their health. Hence Descartes's argument does not follow; see vol. 2, 136.

argued that Descartes is back, at least to some extent, to the troubling possibilities raised in *Meditation* 1. Of course, this is not to say that we need to suppose that the God that exists is as bad or imperfect as the evil daemon; for it could be that He sees that a greater good might be advanced by our being deceived, as Hobbes observed to Descartes. The point is that uncertainty has been introduced. But it is not clear that Descartes recognized this.

10.3.4 Malebranche

However, all this is clearer and more explicit in Descartes's follower, namely, Malebranche, particularly in his Elucidation Six in the *Search after Truth*, on how 'it is difficult to prove that there are bodies'.[10] In short, where Spinoza takes Descartes's idea of the most perfect being in one way or direction – the logical or pantheistic direction – Malebranche goes in the other, explicity allowing that God could be deceiving us, that there is no necessity in the matter. And as with Descartes but even more clearly for Malebranche, what we immediately perceive by our senses is subjective, whereas the real bodies are intelligible. To be sure, for Malebranche we know that since God is infinitely perfect, we believe that He would not deceive us about bodies. But while Malebranche believes this, he is clear that he does not know it – not in the true sense of know, in which he knows that God is a perfect being and that he, Malebranche, exists incorporally. For while he cannot doubt those two propositions, he can doubt that there are material bodies, even that we ourselves have a body that we animate. And as Malebranche says, we believe that there is heat in the fire and light in the sun, but as the new science has shown, we are mistaken in these matters, because our judgements are based on sensory experience. Hence 'our strong propensity to believe that there are bodies', since it too is based largely on sense experience, cannot be known or accepted without qualification. We are free to doubt it.

Malebranche realizes that this sounds mad; but he says he must be truthful – for the sake of truth itself and for those who are struggling to find the truth. But this is not to say that anyone should actually disbelieve in material bodies. There are strong but not necessitating rational grounds to believe in material bodies and, even more compelling grounds from the principles of religion (p. 83). Indeed, as Malebranche says: 'only faith can persuade us that there really are bodies' (pp. 83–84). So for Malebranche as for Descartes there is no cause for real concern or worry in any of this, since we can from philosophy but even more from religion trust that God would not deceive us.

So in the final analysis Malebranche is not at all disturbed or depressed by our inability to know that matter exists. Indeed in at least one place, in his *Dialogues on*

[10] Quotations from Malebranche are from *Malebranche: Philosophical Selections* (Indianapolis: Hackett, 1992), ed. by Steven Nadler; see 77–85.

Metaphysics, he seems positively upbeat about this, since he thinks that it highlights what we do know, namely that the most perfect being, God, exists, and that we exist as non-material being, i.e. souls. For it shows the superiority of the mental over the physical. (p. 215).

10.3.5 Locke

Let us now move from these two rationalists nearer home to Berkeley's great empiricist antecedent: Locke. It is clear that Locke, too, recognized the gap in a very clear and generalized form. Thus in his *Essay concerning Human Understanding* IV. iv. 3, he says:

> it is evident the mind knows not things immediately, but only by the intervention of the ideas it has of them. Our knowledge therefore is real only so far as there is a conformity between our ideas and the reality of things. But what shall be here the criterion? How shall the mind, when it perceives nothing but its own ideas, know that they agree with things themselves?[11]

But while Locke clearly sees the difficulty here – as starkly qua description as Berkeley – he does not seem disturbed by it, either as to the existence of physical things independent of the mind or their correspondence with our ideas. For he thinks that there are 'two sorts of ideas that we may be assured agree with things'. These are simple ideas, which he discusses in the next section and complex ideas, apart from substances, as he says, which he discusses in Section 5. The important ideas, for our purpose here, are the first:

> The first are simple ideas which, since the mind, as has been shown, can by no means make to itself, must necessarily be the product of things operating on the mind in a natural way and producing therein those perceptions which by *the wisdom and will of our maker* [my italics] they are ordained and adapted. From whence it follows that simple ideas are not fictions of our fancies, but the natural and regular productions of things operating without us, really operating upon us, and so carry with them all the conformity which is intended or which our state requires.... (IV.iv.4)

We can see that Locke is really quite close to Descartes here, in that our belief in matter comes from recognizing that we don't cause our sensations, that they are caused by natural things independent of our minds and which he believes have been ordained by God as our state requires.

Locke makes a similar point in *Essay* IV.xi.3, and though he doesn't directly say that he believes in material things because he believes that God wouldn't deceive us, he does bring in the assurance that God provides: '... I think God has given me assurance enough of the existence of things without me: since by their different application, I can produce in myself both pleasure and pain, which is one great concernment of my present state'. And even if we don't know everything of these physical things outside us, still we know enough and have enough assurance from

[11] See Locke, *Essay* (Oxford: Oxford University Press, 1979), ed. by Peter Nidditch, 563.

God so that we can and should 'sit down in a *quiet* [my italics] ignorance of those things ... beyond the reach of our capacities.' (*Essay* I.i.4)

The phrase 'sit down in a quiet ignorance' is noteworthy here, as recalling Berkeley's similar phrase in Section 1 of the Introduction: 'sit down in forlorn scepticism'. So, as in the case of Descartes and Malebranche, Berkeley would not quarrel with the way Locke describes his position as theory. The great difference is in his attitude to it. For instead of enabling us to sit down in quiet ignorance, as Locke thinks, the theoretical position leads, according to Berkeley, to forlorn scepticism.

Now I think we would look in vain in Locke's works for any sense that for him the gap or representationalism has such dire emotional consequences as described by Berkeley, bewilderment, scruples, disturbance and pain. Indeed, as with Malebranche there is even some indication in Locke that there is an up-side to the gap, although not the same up-side as in Malebranche. Whereas for Malebranche, it enhances our sense of the mind's power within the retricted domain of rational psychology and theology, for Locke it can be a useful warning to our pride and excessive belief in our powers of understanding the world. To use Locke's own image, we have a candle, rather than the full light of the sun for getting about in the physical world. And the candle is sufficient for our state. For Berkeley, however, it isn't.

10.3.6 Bayle

So the upshot so far is that we have not been able to find anyone who fits Berkeley's description of the sceptic, who because of representationalism is plunged into an unhappy state of mind. But what of the most famous sceptic of the period, Pierre Bayle? That Bayle is very aware of the sceptical implications of the new science is clear from a number of articles in his *Dictionary*, particularly the article on Pyrrho and the Illustration on it. And it is now generally accepted, since the work of Popkin and Luce, that Bayle was an important influence on Berkeley's development of immaterialism. So Bayle is an obvious candidate to look at.

Unfortunately, there is a serious problem in understanding Bayle, for there has been wide disgreement about Bayle's position, partly arising from the ways he presents his ideas. In many ways he is as evasive and hidden as Kierkagaard, whom he anticipates in many ways. To be sure, the problem of determining his sincerity is one the scholar also has to face with nearly all seventeenth and eighteenth century thinkers, but it is more acute with Bayle.

This is shown in the article on Pyrrho. That Bayle is guarded in the article comes out in the form that he presents his main thoughts, which is an account of a (supposed) dialogue between two Abbes, an account which, Bayle says, 'a very able man' gave him. In this way, he is protecting himself, since he is not claiming that what he is asserting in the dialogue is his own position, or even the unnamed informant present at the dialogue. He is distancing himself. What is clear, however, is that one of the Abbes is drawing heavily on Descartes but especially Malebranche, that since according to the new science God, in effect, deceives us about colour, heat and cold, i.e. secondary qualities, we can't say that God is not also deceiving us about the existence and nature of material things.

In fact, Bayle, or his Abbe spokesman, goes further than Malebranche by arguing, with the help of the Abbe Foucher, that the same reasoning that can prove that we don't know that material things are coloured or hot or cold can also be used to show that we don't know that they are extended, which was, so to speak, the primary quality of Descartes, Malebranche and the Cartesians. So Bayle is undermining even more than Malebranche the reasonable grounds for believing in material objects.

But this is not all. Assuming that the Abbe is correctly pointing out what does follow from Cartesian principles, Bayle is also going further than Malebranche in another way. In order to see this we need to follow his reasoning in detail. After clearly setting out the gap in all its force, he then states: 'I have therefore no good proof of the existence of bodies', referring the reader to Malebranche in a note. He then states that 'The only proof that could be given me of his would be based on the contention that God would be deceiving me if he imprinted on my mind ideas that I have of bodies without there actually being any.'[12] But having stated that this is the only proof, Bayle then goes on to say that 'this proof is very weak', since, if the new science is true, God has been deceiving human beings for thousands of years about the secondary qualities, that fires are hot, etc. So why think He is not deceiving us about matter? Of course, Malebranche had made a similar point, but because he thinks that reason should incline us to belief, he is able to soften this. Whereas Bayle is saying that this is not he case, that reason does not move us in either direction.

So here we surely have someone who sees the full force of scepticism as arising from the dualism of the new science. But the key question is: is Bayle disturbed by this state of affairs? The answer is: not at all. In fact, he is delighted by it. For Bayle's position is that reason has no role to play here. For the belief in matter, as with all other metaphysical matters, is for Bayle entirely a matter of faith and trust in God and his word. So Bayle is even calmer, if anything, than anyone else, since he is not confused by reason. He bases himself entirely on faith. He is a complete fideist, a Kierkagaardian before Kierkagaard, relying entirely on pure faith.

Hence the important point is that, like Sextus, Bayle sees his scepticism as the basis of his calmness, rather than, as Berkeley thinks, drawing him into a worried state of mind. To be sure, there is a difference between Bayle and the Pyrrhonian sceptics. Bayle has a string to his sceptical bow which they lacked. It is not the equipolence of theories and arguments that he seems to rely on and/or that we should accept appearances. For him it is Faith that saves us. For with faith in God and His word, he has an invincible basis for calm: for no reason, however strong, can trump the fideistic stance of 'I believe it because it is absurd or impossible'. As Bayle famously puts it 'Reason should be captivated by faith'. So even if one position can be shown to be reasonably overwhelming and its opposing position can be shown to be self-contradictory, the sceptical fideist has no difficulty in holding the contradictory position or any position based on faith.

[12] Bayle, *Historical and Critical Dictionary*, 197–198.

Thus on the question of matter, Bayle is the antithesis of Hobbes and Spinoza and also, I think, Berkeley, with Descartes, Malebranche, Locke and the great majority of seventeenth and eighteenth century thinkers taking up the middle ground. I say middle ground because their position is that we believe there are material objects partly on reasonable grounds and partly on faith. For Hobbes and Spinoza we know it entirely on reasonable grounds. For Bayle, we believe it (if we do) entirely on grounds of faith or trust.

10.4 Was Berkeley the Unhappy Sceptic?

What then is the upshot to the crucial question I asked and have tried to answer in Section 10.3, namely: who are the unhappy sceptics that Berkeley is adverting to? On the basis of my examination of the most likely pre-1710 candidates, my answer is: I do not think there were any. But then how could Berkeley be so sure of the unhappy effects of sceptical representationalism? Well, one possiblity is that he worked it out logically. He analyzed representationalism and saw that if someone fully understood it, he would inevitably be plunged into unhappiness. But that is like saying that a physican could be convinced that smoking is going to produce a wretched state in smokers before he has actual first-hand empirical evidence that it is so. It seems to me, therefore, that the only plausible answer is that Berkeley knew that representationalism produces the wretched state of mind he describes, because he experienced the wretched state in himself. I think that this is the only feasible answer, even though Berkeley nowhere openly says this, but what he says, for example about Hylas's development and how Philonous came to his position, make sense if that is so.

What I am proposing is that Berkeley was probably the first to experience the dire psychological consequences of the sceptical gap of the new science. Having experienced it, he then sensitized philosophy to it. This is probably the main reason why historians have not seen that Berkeley was the first unhappy sceptic. Their mistake has been to assume anachronistically that it was there before Berkeley, which is understandable because that is how Berkeley presents it. That is, he suggests that there were many philosophers already depressed by the sceptical spectre arising from the gap. Whether he was aware that no one before him seemed to be bothered by the gap, I do not know; although clearly he felt that philosophers should be.

But then how was it that Berkeley alone was bothered? What made him particularly amenable to experiencing the wretchedness of scepticism? I think he gives us the answer in one of his few extant personal notes. It is in the *Philosophical Commentaries*, number 266, where he writes:

> Mem: that I was sceptical [crossed out and replaced by 'distrustful'] at 8 years old and Consequently by nature disposed for these new Doctrines.

The fact that he is able to trace his distrustfulness back to eight is important and fits with his words 'by nature'. What he is saying to himself, I believe, is that his distrustfulness at such an early age shows that it wasn't something accidental or learned. It was a natural or essential part of his character, his true, deep self, we might say. He was not a trusting mind. And we have other evidence that he did not accepts things based on authority or tradition or fashion or consensus; thus in *Philosophical Commentaries*, number 465 he writes:

> + I am young, I am an upstart, I am a pretender, I am vain, very well. I shall Endeavour patiently to bear up under the most lessening, vilifying appellations the pride & rage of man can devise. But one thing, I know, I am not guilty of. I do not pin my faith on the sleeve of any great man. I act not out of prejudice and prepossession. I do not adhere to any opinion because it is an old one, a receiv'd one, a fashionable one, or one that I have spent much time in the study and cultivation of.

From this evidence, I would say that Berkeley was a natural philosopher, someone who had to see or know things with his own eyes or mind, who was not prepared to accept anything at second-hand.[13]

At this point, I think it would be useful if I made a few terminological distinctions. I take it that trust is basically the same as faith, although faith is the term usually used when it is in connection with religion. I also note that distrustful is not the same as sceptical, which I take it is why Berkeley crossed out sceptical and put in distrustful. One can be distrustful without being sceptical. Indeed, being distrustful could well be the surest way of attaining knowledge. But I think natural distrust could lead to scepticism, as I think it did in Berkeley's case, at least for a time. So before he got into philosophy and science, he accepted that there were physical objects out there, independent of his mind. In short, he was, like most of us, some kind of naïve realist, even though we don't formulate it as such, because we haven't yet got into philosophy. It is implicit.

But when at about 16 Berkeley started doing philosophy and science at Trinity College, where the new philosophy and science were being taught, then his implicit commonsense realism came under pressure, either directly from his teachers or through books such as Descartes's *Meditations* and *Principles*, Locke's *Essay*, Malebranche's *Search* and Bayle's *Dictionary*. Then he found himself in a more and more difficult and uncomfortable position. He saw how the gap is problematic. Yet if he was a normally trustful person, he would have gotten over this in the way that most philosophers, as we have seen, did. But because he was *by nature distrustful*, he could not accept that there were material things based on faith or trust in God's goodness. Here his reading of Malebranche and Bayle would have been crucial. For they, particularly Bayle, emphasize how the belief in physical bodies can only be based on faith, that it is not something we know at all. That, I think, would have aroused or activated Berkeley's natural distrustfulness – making the problem of the gap acute for him, plunging him into a state in which he could not be sure one way or the other, and so filling him with a thousand scruples.

And while he could not have liked this unhappy sceptical state of mind, it could well have seemed the most honest and reasonable position, although a painful one.

[13] See my *Berkeley: Experimental Philosophy* (London: Phoenix, 1997) esp. pp. 13–14, and G. A. Johnston, *The Development of Berkeley's Philosophy* (London: Macmillan, 1923), 331–332.

I take it that this fits the evidence I set out above, for example, with what Berkeley says in Section 1 of the Introduction about how philosophy and reasoning can bring about a loss of serenity. But once on this difficult road, he had to find a way of achieving some more workable, calmer, *modus vivendi.* Possibly he did consider the ways of Hobbes and Spinoza, which would recommend themselves to him as at least more reasonable and less trusting than the consensual position. But we need to remember that Hobbes and Spinoza, along with Machiavelli, were regarded as the three devils in the late seventeenth and early eighteenth century – comparable figures for us would be Hitler or Stalin.[14] And Hobbes and Spinoza would also be unattractive not just for their irreligiousness, but also for reasons more intrinsic to their philosophies. Thus Hobbes in rejecting dualism seems to be rejecting the fact of consciousness and subjectivity, which Berkeley later ridicules in various works. For the Hobbist position supposes, as Berkeley says in PHK, Section 142, that our minds are like the objects they perceive. But for Berkeley we do not know mind in the way we know a triangle [for to say so] seems as absurd as if we should hope to see a sound'.[15] And the Spinozist theory, that God is substance or nature that manifests itself deterministically though its dual aspects of thought and extension, would probably have seemed as wild to him as the sceptical dualism it might seem to resolve.

So as I see it, no feasible solution offered itself; so Berkeley had to endure what was probably a painful episode of scepticism, as described in the passages I quoted above in Section 2. That is, until he came on his great discovery, *esse* is *percipi,* the obvious though amazing truth, which overcame the sceptical gap without trust.

Before concluding this essay, I should like to say something about two matters, arising from my account so far. The first is: do we have any direct biographical evidence apart from *Philosophical Commentaries* 266 that Berkeley was likely to be disturbed by the scepticism of the gap? That is, is there any independent evidence that he went through a mentally difficult period in his early philosophical life, which would have been circa 1702–1706, before discovering his *esse* is *percipi* principle? Our difficulty here is that we have so little evidence from Berkeley's early years. The first letter we have from him is from 1706 and the next 1709. But there is some evidence in the first memoir we have of Berkeley, by Oliver Goldsmith, published in 1759–1760. Although it doesn't show that Berkeley suffered from scepticism, Goldsmith's memoir does suggest that Berkeley suffered in a mental way in his early years in Trinity College, at least was undergoing a difficult period in his life.

---

[14] In his *Spinoza Reviv'd* (London: J. Matthews, 1709), William Carroll described Spinoza as 'an eruption from hell', 34.

[15] Also see *Guardian* number 130, where Berkeley says that it is 'plain that no one could mistake Thought for Motion, who knew what Thought was.' Hence Berkeley concludes that the materialistic freethinkers should be regarded as machines or automata and hence spoken of 'in the neuter Gender, using the Term *it* for *him.*' See *The Guardian* (Lexington: The University of Kentucky Press, 1982), ed. by John C. Stephens, 435.

For, according to Goldsmith, having become a student at Trinity College, Berkeley

> soon began to be looked upon as the greatest genius or the greatest dunce in the whole university; those who were but slightly acquainted with him took him for a fool, but those who shared his intimate friendship looked upon him as a prodigy of learning and good nature. Whenever he appeared abroad from his studies which was seldom, he was surrounded by a crowd of the idle and facetious, who followed him not to be improved but to laugh. Of this he frequently complained but there was no redress, the more he fretted he became only the more ridiculous. An action of his however soon made him more truly ridiculous than before...[16]

This action was Berkeley's plan to hang himself, with the help of a college friend called Contarine, in order to experience the sensations preceding death, a plan which he did carry out, according to Goldsmith, and almost lead to his death. Now since Contarine was Goldsmith's uncle, 'from whom [Goldsmith says] he had the story', it seems more than likely that we are getting close to the young Berkeley. And although there is nothing in this story or elsewhere in Goldsmith's memoir that suggests that Berkeley underwent a morbidly skeptical episode or a creative illness, the memoir does paint a picture of a young man who was prepared to engage in extreme behaviour and who appeared eccentric to his fellow students – a picture which, I think, is supported by Berkeley's own self portrait in *Philosophical Commentaries*, number 465, which I quoted above.

The second matter is this. Supposing I am right that the argument for believing in matter which had the most support at the time was that God would not deceive us about such a vital and universal belief, what then was Berkeley's attitude to this argument? If it was the argument of the most significant philosophers of the time, then one would suppose that Berkeley would take it very seriously, since we know he took the gap so seriously. But as far as I am aware, it is not mentioned as such in the PHK, although he responds to 16 objections to his immaterialism and looks at various theories of matter. That in itself is striking. To be sure, he did make up for this omission in the third of the DHP (see p. 243), but there where he does mention it, he dismisses it in less than half a page. (Compare PC number 818). That, I think, is equally striking and puzzling. The thrust of Berkeley's counter-argument is to deny that God has induced mankind to believe in matter, since Berkeley says it is not revealed in Scripture, nor has He made it evident to our natural faculties. Berkeley comes somewhat close to reflecting on this approach in the PHK, Sections 54–57, but in those sections there is no mention of deception or that if matter did not exist then God would be deceiving us. What Berkeley says is that some think that there is a universal belief in matter and hence if that matter does not exist, then we would have to suppose 'the whole world to be mistaken'. Berkeley's criticism here, as in the DHP, is essentially that if matter is taken to be a physical thing existing outside the mind, then only a small number of philosophers

[16] The Goldsmith memoir is reproduced in *George Berkeley: Eighteenth-Century Responses* (New York: Garland Publishing, 1989, ed. by D. Berman, vol. 1, 172.

assert that, and even they can't really be clear about what they are claiming to believe, since it involves a contradiction. Curiously, however, when Berkeley comes (in Section 56) to gives the cause for the apparent universal mistake – which he doesn't really accept as universal – he gives the very reason offered by Descartes and Locke, that our sensory ideas are involuntary and hence must be caused by material objects. So he seems aware of the objection even here, but curiously unwilling to state it in its theological form.

To conclude, then, I have argued that Berkeley went through a painful sceptical episode concerning the existence of the material world, where the choice for him seemed to be between trust in God's goodness, which was the chief response of the most respected philosophers of the time, which Berkeley couldn't accept because he was by nature distrustful, and a more rationally based answer to the problem, but one that seemed to require going the irreligious and speculative ways of either Hobbes or Spinoza – hence a choice between the deep blue sea and the two devils of seventeenth century philosophy. And so, I believe, Berkeley suffered and struggled and finally found the 'obvious and amazing truth' (PC number 279) *esse* is *percipi*, which showed him the way between the devils and the deep blue sea of faith. Because *esse* is *percipi*, there is no gap. With respect to the physical world, appearance is reality, and not, as with the Pyrrhonian sceptics, only what we should accept in a practical way in lieu of reality.[17]

[17] I am grateful to Dr Marek Tomecek for reading an earlier draft of this essay and for providing helpful suggestions on it.

Chapter 11
Berkeley, Spinoza, and Radical Enlightenment

Geneviève Brykman

I put Spinoza in my title in order to signal my paper's purpose. Instead of taking the Enlightenment to be, as it is usually understood, an early-eighteenth-century phenomenon, my intention is to show that Berkeley is to be included in that period of ferment, after Spinoza's death, when the defence of free expression, the critique of religion and of language connected with "mysteries," and the impertinence of any form of authority were openly considered. By contrast, before 1677, the defence of free-thinking was expressed only in clandestine clubs and coteries. Now, in his masterly work *Radical Enlightenment*, Jonathan Israel reveals the central role of Spinoza's philosophy and its diffusion, as early as 1650, in sharpening the human desire for liberty.[1]

One main argument may be put forward in asserting that Berkeley had a share in a radical enlightenment: Spinoza is present, in an equivocal way, in Berkeley's writings. On the one hand, some of his texts exhibit a direct reading of Spinoza, a reading that for this very reason is not a criticism. On the other hand, there are critical points in Berkeley's works too harsh to testify to a direct reading of Spinoza. As Berkeley himself had experienced as early as 1710 (with the publication of the *Treatise of the Principles of Human Knowledge*), many people poke fun at and criticize books without any direct acquaintance with them.[2] Now, in the middle of the seventeenth century, there was already a great deal of prejudice about, and misconstruction of, Spinoza's philosophy. Berkeley had to keep in mind this state of things; and so he did, but in an ambiguous way that induces in his writings an

[1] J.I. Israel, *Radical Enlightenment, Philosophy and the Making of Modernity, 1650–1750*; 810 p., edit. O.U.P., 2001, Preface and ch. 33: *English Deism and Europe*. French translation: J.I. Israel, *Les Lumières radicales, La philosophie, Spinoza et la naissance de la modernité (1650–1750)*, Editions Amsterdam, Paris, 2005.

[2] Letter to Percival, 06 September 1710, in *The Works*, edit. Luce/Jessop (LJ), vol. VIII, p. 36: "I am not at all surprised to find that the name of my book should be entertained with ridicule and contempt by those who never examined what was in it, and want that common justice of trying before they condemn."

G. Brykman (✉)
Professeur émérite, Université Paris-Ouest, Nanterre-La Défense,
70 rue du, Javelot, 75013 Paris, France
e-mail: genevieve.brykman@wanadoo.fr

S. Parigi (ed.), *George Berkeley: Religion and Science in the Age of Enlightenment*, International Archives of the History of Ideas/Archives internationales d'histoire des idées, 201, DOI 10.1007/978-90-481-9243-4_11, "equivocal reading" of Spinoza.[3] Indeed, in Berkeley's works, we can sort his few remarks on Spinoza into two chronologically distinct and strategically very different groups.

First, in the *Philosophical Commentaries* (PC), it is possible to select a group of remarks that give evidence of a direct reading of Spinoza's *Opera Posthuma.*[4] Berkeley sometimes gives a page reference to the 1677 edition of Spinoza's *Opera Posthuma.*

> 826 – Ens, res, aliquid dicuntur termini transcendentales. Spinoza p. 76 prop. 40 Eth. part. 2 gives an odd account of their original. Also of the original of Universals Homo, Canis, etc.
>
> 827– Spinoza (vid. Pref.oper.Posthum) will have God to be Omnium Rerum Causa immanens & to countenance this produces that of St. Paul, in him we live etc. Now this of St. Paul may be explained by my Doctrine as well as Spinoza's or Locke's or Hobbes' or Raphson's.

Secondly, in his published works we can spot Berkeley's official statements on what would better be called *spinozism* than *Spinoza's philosophy.*[5] These official statements may be divided as follows: (a) loose and trite allusions to Spinoza that were commonplaces in the eighteenth century, Spinoza being then pointed to as the spearhead of atheism[6]; (b) from 1732 onwards, particularly in *Alciphron,* more accurate references to Spinoza. In such instances, though still dependent on common negative views of Spinoza, Berkeley's remarks were again indirectly concerned with immaterialism considered as a demonstration of the nonexistence of matter.

Berkeley's direct engagement with Spinoza's works in the *Philosophical Commentaries* makes this manuscript a source of precious information about his views. There we can see that as early as 1708 Berkeley was thoroughly aware of some points at which his and Spinoza's views converged. These points made it urgent to put a clear intellectual distance between himself and Spinoza, the most

[3] On the equivocal reading of authors, see: Leo Strauss, *Persecution and the Art of Writing,* 1952, 1980; *Berkeley's* Alciphron. *English Text and Essays in Interpretation,* edit. 1732, coordinated by Laurent. Jaffro, Geneviève Brykman and Claire Schwartz, Olms-Verlag, 2010.

[4] The main marks of a direct acquaintance with Spinoza's works are in *Notebook A* 622, 625, 824–827, 831, 835.

[5] *Principles* (Pr), 1710, *LJ. I,* §§ 66, 149; *Three Dialogues Between Hylas and Philonous* (DHP), 1713, *LJ.II,* .213–214; *Alciphron* (AL), 1732, *LJ. III,* pp. 163, 281, 324–325; *The Theory of Vision Vindicated and explained* (1733), *LJ. I,* p. 254; *Siris,* § 354, *LJ. V,* p. 160.

[6] See D. Berman, *A History of Atheism in Britain from Hobbes to Russell,* ch. 3–4 , London and New York, 1988; J.I. Israel, *Radical Enlightenment cit,* ch. 33; Geneviève Brykman's *Introd.* to the French edit of *Alciphron,* P.U.F., 1992, vol. III, pp.13–26. In *Alciphron,* Berkeley represents the main free-thinker by the character of Alciphron (a name which has a Greek origin and means "strong head"). This character stands either for Shaftesbury, the Stoics, or Spinoza. Lysicles is sometimes a radical free-thinker and sometimes a radical sceptic, being an eponym for Toland, Mandeville, Collins and Spinoza.

famous freethinker of the period, with whom Berkeley found himself in agreement on at least two basic topics:

1. The monist theory implying that only one infinite substance exists, with finite beings as its modes or effects
2. The strong criticism of language and general ideas

The first similarity will not be surprising; in Berkeley and Spinoza it should be related to the same endeavor of solving fundamental problems left unsolved by Cartesian dualism. In this dualism, the problem of the interaction between two substances ontologically different usually led only to vacuous assertions. In Spinoza and Berkeley, because of an identical purpose of giving a meaning to the word *existence*, and to stress that only what can be conceived of should be talked of, ontological monism is connected to a strict examination of traditional philosophical language.

11.1 Ontological Monism

In the *Philosophical Commentaries* some notes are evidence that Spinoza's philosophy had exerted a fascination on Berkeley. In *Notebook B* we find notes not only about Spinoza, but about all theories that, according to Berkeley, amount either to saying that God is an extended being, or to asserting that extension is an uncreated being independent of God's power.

> 290 – The great danger of making extension exist without the mind, in that if it does it must be acknowledged infinite immutable eternal etc, which will be to make either God extended (which I think dangerous) or an eternal, immutable, infinite, increated being beside God.
>
> 298 – Locke, More, Raphson etc seem to make God extended. 'tis nevertheless of great use to religion to take extension out of our idea of God & put a power in its place. It seems dangerous to suppose extension which is manifestly inert in God.

In *Notebook A*, the working out of immaterialism gives rise to Berkeley's direct reading of Spinoza's *Ethics* and of some of his letters to eminent scholars.[7] To stress the importance of Berkeley's interest in Spinoza's works, we can rely on a comparison with the apparently casual attention that Berkeley paid to Malebranche. After all, Locke had much contributed to spreading in Britain the opinion that Malebranche's doctrine upholding the vision of ideas in God was too subtle to do any harm whatever and so would die of itself.[8]

Going in the same direction, in both *Notebooks*, where Berkeley had no need to be careful and could write without mincing words, a collection of notes indicates

[7] *Notebook A*: 622, 625, 824–827, 831, 835. Spinoza's Letters II, IV, IX, X, XIX, XXI (édit. La Pléiade, Paris, 1954).

[8] Locke, *An Examination of P. Malebranche's Opinion of seeing all Things in God*, in *The Works*, vol. 8, edit.1794, pp. 211–255; see Ch. McCracken, *Malebranche and British Philosophy*, Oxford, 1983, ch. 4 (*Locke's Refutation of Malebranche*), pp. 121–44.

that he thinks that either Malebranche is wrong or he does not prove anything.[9] By contrast, Spinoza's philosophy is an imposing theory that should be overcome urgently. And, at the end of *Notebook A,* Berkeley asserts that he has achieved that goal.

> 824 – My Doctrines rightly understood all that Philosophy of Epicurus, Hobbs, Spinoza etc. which has been a Declared Enemy of Religion comes to the Ground.

To convince himself that it is imperative to refute Spinoza, Berkeley at first says not that Spinoza's philosophy is false, but that it is dangerous. Dangerous as are those doctrines in which there are grounds for thinking of God as an extended being; but dangerous as well on a peculiar score: as Bayle had already noted, Spinoza was a "systematic atheist". Though agreeing on many points with other ancient and modern atheists, Spinoza, according to Bayle, employed a totally new method: the geometrical order of the *Ethics.*[10] Nonetheless, according to Bayle, this method, tied and woven as in geometry, produced a "monstrous hypothesis", contrary to our most evident notions but hard to overthrow[11]:

> It is not as easy to deal with all the difficulties contained in that work [the *Tractatus theologico-politicus*] as to demolish completely the system that appears in his *Opera posthuma,* for this is the most monstrous hypothesis that could be imagined, the most absurd, and the most diametrically opposed to the evident notions of our mind.

Now in *Notebook A,* Berkeley several times remarks that mathematicians possess a wonderful method, but use it on trifles.[12] On the other hand, he remarks that we can look to demonstrations in morals that pave the road to good actions by way of geometry.[13] Berkeley, thus, is not in agreement with Bayle on the means to be used to refute Spinoza. Bayle's target was the 5th proposition of Part I of the *Ethics*: "God is *the only one substance* in Nature, endowed with an infinity of attributes."[14] In such a proposition Berkeley could not possibly see an error; on the contrary, it was a basic metaphysical principle shared among a number of philosophers, of whom he was one.

> 827 – Spinoza (vid.Pref.oper.Posthum) will have God to be Omnium Rerum Causa immanens & to countenance this produces that of St. Paul, in him we live etc. Now this of St. Paul may be explained by my Doctrine as well as Spinoza's or Locke's or Hobbes' or Raphson's etc.

As Bayle had stressed in many places of his *Dictionary,* the principle of God's immanency was an age-long tradition in many philosophies otherwise very

---

[9] Malebranche is wrong: 230, 255, 257, 388, 424a, 548. Malebranche does not prove anything: 265, 288, 358, 424, 686, 686a, 800, 818.

[10] P. Bayle, *Dictionnaire historique et critique,* art. "Spinoza", *Introd.* and *Remark N*; see: Bayle, *Historical and Critical Dictionary, Selections,* ed. by R. Popkin, New York, 1965, p. 288.

[11] *Dictionary,* ibid., pp. 296–297 and *Remark N,* pp. 300–302.

[12] *Notebook A,* 468, 584, 562, 586, 690, 868.

[13] Ibid., 690–691, 853, 858.

[14] *Supra, notes* 12–13.

different.[15] In a letter to Oldenburg, Spinoza asserts that immanency was a doctrine widely held among the ancient philosophers and doubtless by all the Hebrews.[16]

Indeed, Berkeley could rely as well on Locke's *Essay* or on Raphson's *De Spatio Reali*: it is then all the more significant that Spinoza's theory of immanence held his attention more: in Spinoza's philosophy, we live, we move and have our being in God in the proper sense. If, in so far as he was a priest, Berkeley could raise the usual objections against 'spinozism', he could nonetheless, in so far as he was a philosopher, undertake to improve on Spinoza by amending Spinoza's definitions of God and extension.

> 845 – My definition of the word God I think much clearer than that of Descartes & Spinoza viz. ens summe perfectum, & absolute Infinitum or ens constans infinitis attributis quorum unumquodque est infinitum.

In Berkeley's opinion, Descartes' and Spinoza's definitions of God lack the essential divine attributes; they lack what makes Berkeley's monism altogether different from Spinoza's. According to Berkeley, the infinite power of God is not blind, but is to be conceived as a providential eternal activity, watching over the universe and keeping it in order at every moment.[17]

> 812 – The properties of all things are in God i.e. there is in the Deity Understanding as well as Will. He is no Blind agent & in truth a blind agent is a contradiction.

As to extension, contrary to what Spinoza wrote to Oldenburg, Berkeley notes that it cannot be conceived without something else.

> 844 – Dico quod extensio non concipitur in se et per se contra quam dicit Spinoza in ep. I ad Oldenburgium.

At the end of the *Philosophical Commentaries*, a basic difficulty was not yet solved: extension cannot be conceived as independent of God, yet it is not a mode or property in God.[18]

11.2 The Criticism of Language and General Ideas

In the *Philosophical Commentaries*, and in the later works, Berkeley's agreement with Spinoza about the oneness of being is joined to another similarity that, though apparently more limited, is no less decisive for the cogency of Berkeley's monism. This second agreement lies in exhibiting linguistic traps that result from general

[15] *Dictionary*, articles: "Anaxagore", "Epicure", "Pauliciens", "Spinoza", Remark.

[16] Letter LXXII: "I assert, as St. Paul and nearly all ancient philosophers said, though in another way, that all things are in God and move in God; and I dare even say that this assement was to be found in all ancient Hebrews."

[17] *Principles*, 45–48, 60–62, 146–156; *DHP2*, *LJ II*, pp. 210–213.

[18] *Notebook A*, 878, 886; pr., 49.

ideas that some take to be "abstracted" by a fine intellectual operation, though they are in fact the blind mechanical result of imagination.

If we judge by a number of notes in *Notebooks B* and *A*, Berkeley may have thoroughly subscribed to Spinoza's statement[19]:

> Surely, human affairs would be far happier if the power in men to be silent were the same as that to speak. But experience more than sufficiently teach that men govern nothing with more difficulty than their tongue, and can moderate their desires more easily than their tongue.

In a number of Berkeley's notes, similar judgments are to be found, constantly vindicating silence, which silence consists first in holding one's tongue.[20] For instance:

> 553 – I must not say that the words *thing, substance,* etc have been the cause of mistakes, but the not reflecting on their meaning. I will be still for retaining the words. I only desire that men would think before they speak & settle the meaning of their words.

> 686 – Scripture & possibility are the only proof with Malbranch add to these what he calls a great propension to think so. this perhaps may be question'd. Perhaps men if they think before they speak will not be found so thoroughly persuaded of the existence of matter.

In the *Principles* and *Three Dialogues,* the critical examination of language is ostensibly present and mainly refers to the demonstration of immaterialism. But in the *Philosophical Commentaries* such a critical examination is not limited to material substance; it is dangerously applied to any substance or words like 'thinking thing', 'will', or 'liberty'. Indeed, Berkeley's distrust of language led him far from a submission to Scripture.

> 577 – The very existence of ideas constitutes the soul.

> 580 – Mind is a congeries of perceptions. Take away Perceptions & you take away the Mind put the Perceptions & you put the mind.

> 581 – Say you the Mind is not the Perceptions but that think which perceives. I answer you are abused by the words *that* and *thing* these are empty words without a meaning.

> 615 – The Will not distinct from Particular volitions.

> 627 – We are imposed by the words, *will, determine, agent, free, can,* etc.

> 631 – You tell me according to my Doctrine a Man is not free. I shall answer, tell me what you mean by the word free & I shall resolve you.

> 637 – Say you there must be a thinking substance. Something unknown which perceives & supports & ties together the ideas. Say I, may it appear there is any need of it & you shall it for me. I care not to take away anything I can see the least reason to think should exist.

---

[19] *Ethics,* Part. III, pr. 2, *scholium.*

[20] On similarities between Berkeley and Wittgenstein, concerning the basic principle: "What we cannot speak about we must pass over in silence" (*Tractatus Logico-philosophicus,* trad. Pears, London, 1974, p. 74), see Geneviève Brykman, *Berkeley et le voile des mots,* Paris, Vrin, 1993, pp. 202–214.

In *Notebook A,* these notes are to be found shortly before several notes bearing the marks of Berkeley's reading of Descartes' *Meditations* and Hobbes' *Objections,*[21] as well as of a particular attention to Spinoza's *Ethics* and *Letters.*[22] Like Spinoza, Berkeley may have thought that the lexicon about liberty was only a shelter for our ignorance about the nature of the union between body and soul – that is, the interaction of two ontologically different substances. Berkeley's irony about the pineal gland in Descartes has no equal, save only Spinoza's own sarcastic remarks in the *Ethics*[23]: some men, not knowing how what they call the "will" might move their body, nonetheless pretend to explain it; to this end they concoct a seat for the soul "that is really just a sign and refuge of their ignorance; they usually excite laughter or loathing."[24] In the *Guardian,* Berkeley was on the side of laughter.

With the critical examination of the words *will* and *liberty,* and the refutation of Cartesian dualism, Spinoza had shown Berkeley the right way to get rid of *material substance.* In both cases there is at first a kind of natural illusion, but such an illusion is not a mistake. It becomes a mistake only when men, using language without reflecting on what they say, maintain this illusion and strengthen it. Thus, according to Spinoza, so far as men believe that they are free when they are conscious of their desires but are unaware of the causes by which they are determined to have those desires, there is no mistake.[25] But, concerning what men usually assert – that human actions depend on the decisions of their wills – these statements are only words without any reference.[26] Berkeley will argue in the same way about the illusion of distance in the external world.[27]

---

[21] See the entries 779–780, 782, 784–786, 790, 794–805, 816–823, 841–848.

[22] Ibid., 824–827, 831, 844–845.

[23] In a pair of articles in the *Guardian,* Berkeley imagines that he was furnished with a snuff «which, if taken in a certain quantity, would not fail to disengage [his] soul from [his] body. Then, his soul was at liberty to go wherever she pleases and to enter the pineal gland of learned philosophers and men of pleasure" (*LJ VII,* pp. 186–187).

[24] *Ethics,* Part *I, Appendix;* Part *II, prop.* 35, *schol.*; *V, Preface:* "This opinion [...about a complete command over our emotions] is not a little favoured by Descartes, for he held that the soul or mind is particularly united to a certain part of the brain, called the pineal gland, by means of which it feels all the movements that take place in the body and external objects and which the mind, by the very fact that it wishes, can move in various ways. [...] I cannot sufficiently wonder that a philosopher, who clearly stated that he would deduce nothing save from self – evident bases of argument, and that he would assert nothing save what he perceived clearly and distinctly – one, moreover, who reproves the Schoolmen for wishing to explain obscure things by means of occult qualities, should take an hypothesis far more occult than all the occult qualities."

[25] *Ethics, Part I, Appendix.*

[26] *Ethics, Part II, prop.* 35, *scholium*: "For that which they say, that human actions depend on the will, are words they do not fathom."

[27] *An Essay Toward a New Theory of Vision (NTV,* 1709). On the natural illusion of a perception of distance by sight, see §§ 41, 43–48, 51, 64; Berkeley stressed it in § 74, asking: "what is it can put this cheat in the understanding?"; on the quick and sudden suggestions of imagination: §§ 51, 66, 95, 126,145–146, 159.

The illusion of visual distance is not present at the very moment a child is born, or when a man born blind but enabled to see opens his eyes on the world for the first time:

> From what has been premised", Berkeley says, " it is a manifest consequence that a man born blind, being made to see, would, at first, have no idea of distance by sight; the sun, and stars, the remotest objects as well as the nearer, would all seem to be in his eye, or rather in his mind. The objects introduced by sight would seem (as in truth they are) no other than a new set of thoughts or sensations, each whereof is as near to him as the perceptions of pain and pleasure, or the most inward passions of the soul.[28]

The illusion starts to insinuate itself at the second glance, one might say, because of the quick and imperceptible suggestions of imagination taking visual ideas as signs of tactual ideas, thoroughly heterogeneous from them.[29]

> Light and colours are allowed by all to constitute a sort or species entirely different from the ideas of touch; nor will any man, I presume, say they can make them perceived by the same sense; but there is no other immediate object of sight beside light and colours. It is therefore a direct consequence that there is no idea common to both senses.[30]

This illusion becomes more and more a prejudice as it is steadfastly strengthened through language.[31]

> No sooner do we hear the words of a familiar language pronounced in our ears, but the ideas corresponding thereto present themselves to our mind: in the very same instant the sound and the meaning enter the understanding, so closely are they united that it is not in our power to keep out the one, except we exclude the other also. We even act as if in all respect we heard the very thoughts themselves. So likewise the secondary objects, or those which are only suggested by sight, do often more affect us, and are more regarded than the proper objects of that sense. [...] Hence it is we find it so difficult to discriminate immediate and mediate objects of sight. They are, as it were, most closely twisted, blended, and incorporated together. And the prejudice is confirmed and riveted in our thoughts by a long track of time, by the use of language, and want of reflexion.[32]

> The prejudice becomes an error as soon as men are not satisfied merely to act as if matter existed, but build on theories that suppose that a material substance is the cause of sensible ideas.[33]

> *Philonous*: I shall never quarrel with you for an expression. *Matter*, or *material substance*, are terms introduced by philosophers; and as used by them, imply a sort of

[28] *NTV*, § 41.

[29] *NTV*, §§ 49, 61, 97, 111, 121, 129, 139, 147–148, 152; Berkeley sometimes describes the order between visual and tactual ideas, as Spinoza does for the common order between attributes: see § 111 and *DHP 3*, *LJ II*, 241: "This connexion of sensations with corporeal motions means no more than a correspondence, in the order of Nature, between two sets of ideas or things immediately perceivable."

[30] Ibid., § 129.

[31] *NTV*, §§ 51, 61, 64, 94–95, 110, 120,140–147, 152.

[32] Ibid., § 51.

[33] *Principles*, §§ 52–57, 72–75, 108–110, 150–153; *DHP3*, *LJ2*, 261–262.

> independency, or a subsistence distinct from being perceived by a mind: but are never uses by common people; or if ever, it is to signify the immediate objects of sense.[34]

In *Book III* of his *Essay concerning Human Understanding*, Locke already stressed that many human errors and prejudices come from language. Nonetheless, Locke seemed to share the widespread opinion that words signified abstract general ideas.[35] Now Berkeley, like Spinoza, firmly asserts that general ideas are neither explained nor justified by abstraction.[36] The merciless censure that Spinoza directed at jargon, showing what makes us take *humanity* as the cause of Peter and Paul, or *liberty* as the cause of such-and-such an action,[37] undoubtedly brought a critical leaven to Berkeley's *Introduction* to the *Principles*.

Locke conceived abstraction as an intellectual process by which the mind omits from an idea the circumstances and details that may determine it to a particular existence, so that only the essential characteristics are kept, thereby making the idea a general one.[38]

> Words become general by being made the signs of general ideas; and ideas become general by separating from them the circumstances of time and place and any other ideas that may determine them to this or that particular existence. By this way of abstraction they are made capable of representing more individuals than one: each of which, having in it a conformity to that abstract idea, is (as we call it) of that sort.[39]

As for Spinoza, he described abstraction not as a kind of elimination, but as the result of a confused piling up of sensory experiences; not as an intellectual process, but as the mechanical achievement of blind imagination.[40] According to Spinoza, the terms called *transcendentals*, which were traditionally thought to hold the highest place in the hierarchy of general ideas, are only the highest confusion of ideas.

> Lest I should omit anything that is necessary to be known, I shall briefly add the causes from which the terms called transcendental have taken their origin, such as *being* or *something*. These terms have arisen from the fact that the human body, since it is limited, is only capable of distinctly forming in itself a certain number of images (I have explained what is an image in II, prop. 17, *scholium*); and if more than this number are formed, the images begin to be confused [...]. And from similar causes have arisen those notions that are called universals or general, such as *man*, *dog*, *horse*, etc. I mean so many images of men arise in the human body e.g., so many images of men are formed at the same time, that they overcome the power of imagining, not altogether indeed, but to such an extent that the mind cannot imagine the small differences between individuals (e.g., colour, size,

[34] *LJ II*, p. 261.

[35] Locke, *Essay*, *Book*, II, ch. 11, §§ 9–11; *Book*, III, ch. 3, §§ 6–14.

[36] *Principles*, *Introd.*, *LJ*, *II*, 25–40; *Manuscript Introd.*, edition diplomatica by B. Belfrage, Oxford, 1987.

[37] *Ethics*, II, 49, *scholium* 2.

[38] *Essay*, *Book III*, ch. 3.

[39] *Essay*, *III*, ch. 3, § 6.

[40] *Ethics*, *II*, prop.17–18; prop. 40, *scholium* 1.

etc.) and their fixed number, and only that in which all agree, in so far as the body is affected by them.[41]

Hence in *Notebook A,* Berkeley's attention to Spinoza's statements on this particular topic:

826 – Ens, res, aliquid dicuntur termini transcendentales, Spinoza p. 76, prop. 40, Eth. Part. 2 gives an odd account of their original. Also of the original of all Universals, Homo, Canis, etc.

Berkeley judges Spinoza's account "odd" in that it turns imagination into a mechanical process in the brain, something that immaterialism would not admit.

11.3 The Human Mind as an Incorporeal Active Substance

By contrast, according to Berkeley, imagination is an active power that he goes so far as to conceive of as a human prerogative: it is imagination that makes us images of God, because it is in us a mark of a creative power and of the presence of liberty. At the end of *Notebook A,* Berkeley insists on that creative power of imagination, which is a basic sign of spirit as an active substance.

753 – Qu: whether composition of ideas be not that faculty which chiefly serves to discriminates us from brutes. I question whether a brute does or can imagine a blue horse or chimera.

830 – Why may we not conceive it possible for God to create things out of nothing. Certainly we ourselves create in some wise whenever we imagine.

Moreover, to defend the very notion of creation, note 831 exhibits what is to be the bone of contention between Berkeley and Spinoza:

831 – *Ex nihilo nihil fit* (says Spinoza Op. posth. p. 464) & the like are called veritates aeternae because nullam fidem habent extra mentem. To make this axiom have a positive signification, one should express it thus. Every idea has a cause i.e. is produced by a will.

Indeed, the conception of God as pure Act and the admission of a will in human beings are, in Berkeley, the essential lines in his opposition to Spinoza.[42] By the end of the two *Notebooks,* ontological monism was to be joined both to the peculiar requirements of immaterialism and to the necessity of being in accordance with common sense and the Anglican Church, each requiring that liberty and responsibility be upheld. Accordingly, Berkeley's problem was to keep a dualism

[41] *Ethics, II,* prop.40, *scholium* 1.

[42] *Notebook A,* 812: "The properties of all things are in God I.E. there is in the Deity Understanding as well as Will. He is no Blind Agent & in truth a blind agent is a contradiction". Spinoza, *Ethics, II,* prop. 43, *scholium*; prop. 48–49 and *scholium.*

(between an active spirit and passive ideas) in a metaphysics that, by the suppression of matter, was basically monist. Here was a conceptual tension, and even a contradiction; because Berkeley's monism requires *changing things into ideas,* but dualism requires *changing ideas into things.*[43] To this irreducible question Berkeley seems to have replied as Descartes did to Elisabeth: it is wiser to live than to speculate.[44]

11.4 Conclusion

In 1710, in the *Principles,* Berkeley states that, in order to struggle against the assaults of skepticism, nothing is more important in erecting a firm system of knowledge than to begin with a clear explication of terms like *thing, reality, existence.*[45] In 1713, from the lips of Philonous, he stresses that matter is of no use in making Creation intelligible.[46] Nonetheless, by the end of his conversation with Hylas, Philonous admits that the word *matter* may be kept in ordinary life and everyday language.[47] Lastly, in 1732, in *Alciphron,* Berkeley calls upon his character Crito to defend Christianity, by saying: "The more room for doubt the more room there is for faith."[48] At the end of *Alciphron,* when Dion asserts that he had it by hearsay that Spinoza is "a man of close argument and demonstration," Crito answers:

> He did demonstrate; but it was after such a manner as anyone may demonstrate anything. Allow a man the privilege to make his own definitions of common words, and it will be no hard matter for him to infer conclusions which in one sense shall be true and in another false, at once seeming paradoxes and manifest truisms.[49]

Making one's own definitions of common words? That is exactly what Berkeley was reproached for doing in 1710–1713 for common words like *thing, cause, idea,* and, above all, *existence.* To defend religion against the attacks of skeptics, the rise of deism and materialism, Berkeley would rather put forward the malleability and efficiency of the ordinary language, than the explicit definitions of sciences and philosophy. Thus, though careful and in some way fascinated by the scientific

---

[43] *DHP3, LJ2,* 244: *Philonous*: "I am not for changing things into ideas, but rather ideas into things".

[44] Descartes to Elisabeth, June 28, 1643.

[45] Berkeley, *Principles,* § 89.

[46] *DHP3, LJ II,* pp. 250–57; Letter to Percival, *LJ VIII,* p. 37.

[47] *DHP3, LJ II,* pp. 261–262.

[48] *Alciphron,* 7, *LJ III,* p. 322.

[49] *Alc.*7, p. 324.

advances of his time, Berkeley is representative of a broad stream that, at the beginning of the eighteenth century, made futile attempts to turn back from the Radical Enlightenment. And it is from his own attempts that, in *Alciphron*, Berkeley can say that some scholars considered Spinoza "the great leader of our modern infidels."[50]

[50] Ibid., 324. This statement is made by Criton, who states that, in Spinoza's works, "are to be found many schemes and notions much admired and followed of late years: such as undermining religion under the pretence of vindicating and explain it; the maintaining that it is not necessary to believe in Christ according to the flesh; the persuading men that miracles are to be understood only in a spiritual and allegorical sense; that vice is not so bad a thing as we are apt to think; that men are machines impelled by a fatal necessity". I should like to express my gratitude to Katherine and Charles McCracken for their careful examination of the English version of my text and for their expert advices concerning the transposition of Berkeley's quotations.

Chapter 12
Was Berkeley a Spinozist? A Historiographical Answer (1718–1751)

Caterina Menichelli

"By the time of Immanuel Kant, Berkeley had been called, among other things, a sceptic, an atheist, a solipsist and an idealist. In our own day, however, the suggestion has been advanced that Berkeley is better understood if interpreted as a realist and man of common sense".[1] In his book of 1965, H. M. Bracken showed with this sentence the different ways in which critics and researchers, since the second half of 1700 to the present, have defined the philosophy of George Berkeley; the variety of these adjectives appears surprising and, at the same time, it suggests very interesting areas for research which, far from being conventional labelling, aim to closely investigate and to revitalise the interest in Berkeleian thought.

Owing to the limits of space, we cannot in this paper completely and exhaustively study all the interpretations of Berkeleian philosophy.[2] The enquiry, therefore, will concern a specific chapter of Berkeley's historiography; specifically, it will consider a group of documents that, within the ambit of the *early receptions*, show the presence of a controversial interpretative current, which seems to have been all but forgotten by modern critics. Between 1718 and 1751, indeed, Berkeley was explicitly accused of being a Spinozist and even an atheist. These accusations were evidently very serious, above all for a philosopher like Berkeley, who considered himself as a paladin of faith and a defender of religion against the attacks of atheists and free-thinkers.

---

[1] Harry M. Bracken, *The Early Reception of Berkeley's Immaterialism. 1710–1733* (The Hague: Nijhoff, 1965), IX.

[2] George Berkeley's works quoted in the text are: *A Treatise concerning the Principles of Human Knowledge* (TK), *An Essay Towards a New Theory of Vision* (NTV), *Three Dialogues Between Hylas and Philonous* (DHP), *De Motu* (DM), *The Theory of Vision ... Vindicated and Explained* (TVV), *Alciphron, or the Minute Philosopher* (*Alc*); *Siris* (*Siris*); *Notebooks*, also known as the *Philosophical Commentaries* (PC). Main English edition: George Berkeley, *The Works of George Berkeley, Bishop of Cloyne*. 9 vols. Eds A. A. Luce and T. E. Jessop. London: Thomas Nelson & Sons, 1948–57.

C. Menichelli (✉)
Località Seola Alta, n. 13, 62032, Camerino (Macerata), Italy
e-mail: catemen@libero.it

S. Parigi (ed.), *George Berkeley: Religion and Science in the Age of Enlightenment*, 171
International Archives of the History of Ideas/Archives internationales d'histoire des idées 201,
DOI 10.1007/978-90-481-9243-4_12,

As you will see, not all the texts that I am going to consider are similar; they differ in particular in the complexity of the topics discussed in them, ranging from a brief affirmation, in the *Acta Eruditorum*, concerning the presence of a Spinozistic source in the *Three Dialogues Between Hylas and Philonous* (DHP), to the complex and interesting accusations that Andrew Baxter, Andrew Ramsay and Robert Clayton made in their books.

My aim is to focus on some interpretative models which, though in a minority among the traditional interpretations of Berkeley's philosophy, had an important and not marginal role in the first phase of his reception.

12.1 Critics and Reviewers: 1711–1727

The very first reviews of the *Treatise Concerning the Principles of Human Knowledge* (TK; *Journal des Sçavans*, 1711), DHP (*Journal Litéraire*, 1713) and both these (*Mémoires de Trévoux*, May and December 1713) show the beginnings of a kind of accusation which, some years later, became more direct and severe; although hidden under a veil of curious strangeness, derived from the denial of the existence of external objects, the rejection of matter is always present and it generates insidious doubts. Berkeleian works do indeed not belie the increasingly worried analyses of the reviewers concerning the existence of a cosmos which, deprived of the material substratum, becomes a monistic–spiritualistic reality, barely discernible from the First Cause.

In 1718, the French Jesuit René-Joseph Tournemine, the founder of *Mémoires de Trévoux*, attacked Berkeley in his preface to the book of François de Salignac de la Mothe Fénelon: *Traité de l'Existence et des Attributs de Dieu.*[3] In Fénelon's *Oeuvres Philosophiques* (published by Andrew Ramsay[4] in 1718), the section *Refutation of Spinoza* gave Tournemine the opportunity of attacking the Spinozistic heresy. Although, in Bracken's opinion, there was no doubt that the proper target was Malebranche, Berkeley was the only philosopher explicitly quoted.

In this particular passage, Tournemine indicates one of the essential and more problematic questions of Berkeleian doctrine: "Le livre Anglais d'un certain

---

[3] Fénelon, archbishop of Cambrai, exerted an important role in French philosophy and culture and his influence was extended until the XVIIIth century. Fénelon's writings excited the worried reaction of Tournemine: in effect, in the editions of 1712 and 1713 of *Démonstration de l'Existence de Dieu*, he had underlined some elements dangerously close to Malebranchian theory. For this reason he tried to correct and mitigate them, adding a preface with the title *Réflexion sur l'Atheisme* to the 1713 edition.

[4] Ramsay was for a long time Fénelon's personal secretary. In his work *Philosophical Principles unfolded in a geometrical order* (1748) there is one of the most important and crucial accusations of Spinozism concerning Berkeley. It is possible to get more information about the relationship between Ramsay and Fénelon in Albert Cherel, *Un Aventurieur Religieux au XVIIIe Siècle. Andrew Michael Ramsay* (Paris: Libraire Académique Perrin e C., 1926).

Berkeley a rendu publics ces nouveaux efforts de l'Incrédulité. Les impies de cette Secte ne disent plus que tout est matière, ils disent que tout est esprit; le Monde, selon eux, n'est composé que d'êtres pensants."[5] He is probably referring to a presence in this philosophy of a sort of inclination to idealism, that is to say a theory which only uses spirit in order to explain the Universe.

In his preface, Tournemine specifies three topics concerning this new philosophy (even if it was only in order to emphasize "*tout le ridicule*" of it); the first one is relating to the foundation of idealism. He wondered, first of all, if it is founded on a certain principle or on a kind of experimental evidence; the author's answer is unequivocal: "*Non: elle se fonde uniquement sur des conjectures opposées au sens commun, sur des suppositions chimériques, sur des possibilités au moins douteuses.*"[6] The second point is a sort of concession to Berkeley's philosophy; initially, the French author wanted to back up his proposal not to consider external objects as deceptive phantoms (because they are deprived of matter), but to consider them as real beings, which preserved all their sensible qualities. Provided that this principle is true, Berkeley is right when he argues that immaterialism does not reduce the strength of the physico-theological proof of God's existence: "On n'évite point dans ce Système la preuve de l'existence de Dieu, tirée de l'excellence de ses ouvrages: cette excellence, cette perfection qui décèle l'ouvrier, ces marques de sa sagesse infinie, ces arguments démonstratifs de son excellence; je les trouves dans mes pensées [...]."[7] Tournemine concludes his reflections concerning the atheism of immaterialists. He wonders what the essence of thinking beings is, which, according to this doctrine, compounds the world; and, above all, he wonders again if they existed from eternity or they started to exist at a precise moment in time; according to Tournemine, all the answers to these questions give absurd results for any rational being. He ends: "Ils regarderont sûrement les Immatérialistes comme ces hommes qui s'imaginant être de verre, craignait à tout moment d'être cassé."

The brief comment, concerning the 1725 edition of DHP, which appeared in *Acta Eruditorum*,[8] consists of a very short Latin text. Berkeley's philosophy is considered as a "mixture" of three doctrines: those of Malebranche, Descartes and Spinoza. "*Ita Berkeleius paradoxon suum de non existentia materiae speciose satis defendit, de cujus veritate alii judicent; de origine, quicquid Autor dissimulet, sic sentimus, ex Cartesii, Malebranchii & Spinosae philosophiarum mixtura prognatum hoc γιβιχον θηριον.*"[9] The reviewer judges the attempt to deny the existence of matter to be a paradox, even if Berkeley tried to give it a rational justification. The reviewer does not closely investigate the doctrine, but he believes he knows its

[5] Bracken, *The Early Reception*, 109.

[6] Ibid., 110.

[7] Ibid.

[8] *Acta Eruditorum*, Leipzig, 1727: 379–383.

[9] "Thus Berkeley gives a fairly plausible defence of his paradox of the non existence of matter. Whether it is true or not I leave others to judge; but as for its origin this Lycean Beast seems to me, despite the author's dissembling, to have a sprung from a mingling of the philosophies of Descartes, Malebranche and Spinoza." (Latin text, Ibid., 1).

sources (in spite of Berkeley's efforts at dissimulation): not only Descartes and Malebranche, but also Spinoza are quoted.

It is worth remembering that Berkeley had the precise intention of fighting against Spinoza and his impious system,[10] which was guilty of having put matter directly among the attributes of God, contaminating his essence. In this way the Amsterdam philosopher reduced the power of God's presence in the world, substituting for the good and generous action of the Spirit an infinite series of causes and effects that are dominated by blind necessity.[11]

One fact is clear: a group of philosophers and commentators identified, in Berkeleian immaterialism, if not properly an explicit conformity with Spinoza, at least a sort of sliding toward pantheistic results. Even if Berkeley declared a profound aversion to this doctrine, which destroys the basis of the religious system that he wanted to defend, this does not *de iure* rule out the existence of reciprocal exchanges and veiled influences.

12.2 Andrew Baxter

In the second volume of his book *An Enquiry into the Nature of the Human Soul* (1733),[12] Andrew Baxter places a chapter expressly dedicated to the confutation of Berkeleian immaterialism; here Baxter accuses Berkeley of sliding in the direction of atheistic positions: "I might also mention the direct tendency of this improvement to Atheism."[13] Why did the author pronounce such a severe accusation against a Christian Bishop, without taking into account Berkeleian declarations concerning his intention of defending religion from the attacks of immaterialism and free-thinking? According to Baxter, from Berkeley's works one can deduce, first of all, that no one can be sure of other men's existence, of their souls or of their liberty: "Besides since no man can be certain of the existence of other men upon this scheme; [...] taking away the existence of their bodies there is no kind of evidence left for the existence of the souls of men, who by the abuse of their freedom might tempt us."

---

[10] It is possible to find this opinion in the young Berkeley's notebooks (see PC, *Notebook A, entries* nn. 825, 844, 845).

[11] In the Berkeleian works there are not many places in which Spinoza is quoted, but those rare examples have an unequivocal accusatory tone, as for example, this: "[...] those wild imaginations of Vanini, Hobbes and Spinoza, in a word the whole system of atheism is it not entirely overthrown by this single reflection on the repugnancy included in supposing the whole, or any part, even the most rude and shapeless of the visible world to exist without a mind?" (George Berkeley, *Three Dialogues between Hylas and Philonous* (Second Dialogue), *The Works,* vol. II, 213).

[12] Andrew Baxter, *An Enquiry into the Nature of the Human Soul,* 2 vols. (London: Millar, 1733, 3rd ed., 1745).

[13] Ibid., 291.

It seems that God becomes alone responsible for our actions and sensations: "and since it is said that God excites in us all the ideas, which we fancy are excited by bodies." The logical (albeit dangerous) conclusion from a philosophy in which the Infinite Mind is considered the only source of the Universe's life and action, is that this same Mind is accountable even for sin and transgression.

The principal aim of the *Enquiry* is to demonstrate, in an irrefutable way, the existence of a spiritual substance and its immateriality. But, if we accept the Berkeleian principle according to which only the things that are perceived by our sensitivity can really exist, there will be no perception of matter nor any proof of the soul's existence (because not even the soul is perceivable): "Now (to observe here the extent of this kind of doubting) this argument will equally show spiritual substance to be a contradiction in terms as well as matter: for we are percipient of nothing but our own perceptions and ideas, with respect to the soul of another man as well as with respect to his body."[14]

If we extend this killing blow, it will count even in the case of God: "it equally demonstrates all substance out of existence save the mind thus percipient without excepting the Deity himself."[15] Baxter, therefore, not only justifies the accusations of "pyrrhonism" and "egoism" against Berkeley, but he also decidedly condemns his philosophical process, judging it to be inconsistent and contrary to religion: "it is not easy to guess what justifiable design a man could propose [...] as to demonstrate that the beautiful system of material nature [...] their usefulness to mankind and the kindness of God in bestowing them are nothing but a dream within the mind."[16]

But Baxter probably hoped to reach another and deeper result by this accusation. Indeed, he wanted to demonstrate, in effect, that two elements are essential for the constitution of the Universe[17]: God (the active principle) and matter[18] (the passive substratum, which is able to receive any shape or action). If we admit one of these, we will have to do the same, inevitably, with the other: "The necessity of an immaterial powerful Being who first made this dead substance matter, originally impressed, and still continues to impress motion upon it. The first thing that appears in his nature as he is thus discovered is his immateriality, being the powerful

---

[14] Ibid., vol. II, 258.

[15] Ibid.

[16] Ibid., vol. II, 248.

[17] Also Bracken underlined this aspect: "He sought to argue that God is the continuing source of all action and cohesion in the universe and that the evidence for His existence comes from the material world" (Bracken, *The Early Reception*, 34–35).

[18] Yet J. W. Yolton emphasised that Baxter wants to demonstrate both that God is uniquely responsible for action and movement and that there exists a being that is completely deprived of activity, but ready to receive forms, like matter. J. W. Yolton, *Thinking Matter. Materialism in Eighteenth-Century Britain* (Oxford: Blackwell, 1983), 95–96.

Creator and Mover of matter; for it is already evident, and shall still be made more plain that such a powerless, dead substance, as matter, must owe its existence to something else."[19] Using an effective metaphor, Baxter argues that it would be as if a musician did not have his instrument on which to play: "What would one conclude [...] if he sometimes saw a Musician play on an instrument, [...] the sensory is the instrument which is sometimes moved [...]."[20]

According to the *Enquiry*'s author, the existence of matter is plainly evident because, if it did not exist, no one would have its idea in his mind, or (as Tipton and McCracken claim) the same idea would be inconsistent and contradictory.[21] The fact that we think of matter makes the Berkeleian argument ineffective. Matter is nor perpetual or uncaused, and it cannot be the eternal effect of an eternal cause; rather it should have been created in the same moment in which the First Principle started to exercise his power: "It must have been created (and out of nothing too: not certainly out of a half-finished phantom of a substratum) when the power of this immaterial cause was first exerted, to make it a solid resisting substance."[22] In the same way it has to keep existing like an inert substratum, which obeys the infinite action of God.[23]

On the other hand, if we did not accept the existence of both substances, we could slide toward equally false opposite theoretical positions. Admitting only the material principle, we would generate a philosophy infused with the most blind materialism and mechanism, like Hobbes'. In opposition to that – Baxter observes in the *Enquiry* – Berkeley elaborated an idealistic doctrine, in which matter has definitely disappeared and things have been reduced to mere illusions:

> For it seems impossible that a man should be seriously persuaded that he has neither country nor parents, nor any material body, nor eats nor drinks [...] but that all these things are mere illusions, and have no existence but in the fancy. [...] if there be nothing but ideas instead of the objects of our ideas [...] therefore must be impossible.[24]

If the Irish philosopher believed he had won the battle against atheists and sceptics, he was wrong, because he did not see that he had favoured them. Baxter's last observation is caustic: Berkeley's behaviour looks like the conduct of some women who, wanting to silence the defamatory insinuations concerning their reputation, become prostitutes: "This is I think, as if one should advance, that the best way for a woman to silence those, who may attack her reputation, is to turn a common prostitute. He puts us into a way of denying all things, that we may get rid of the absurdity of those who deny some things."[25]

---

[19] Baxter, *An Enquiry*, vol. I, 80.

[20] Ibid., vol. II, 43.

[21] McCracken-Tipton, *Background Source Materials*, 195.

[22] Baxter, *An Enquiry*, vol. I, 325–327.

[23] "This itself appeared to be the power of this immaterial Cause indefinitely impressed upon, and exerted in every possible part of matter". Ibid., 322.

[24] Ibid., 239–240.

[25] Ibid., 262–263.

12.3 Andrew Michael Ramsay

The most important philosopher who severely and rigorously accused Berkeley of Spinozism was Andrew Michael Ramsay[26] in his book *The Philosophical Principles of Natural and Revealed Religion unfolded in a Geometric Order* (1748).

Born in 1686 in the little Scottish town of Ayr, he published in 1727 *Les Voyages de Cyrus* (followed by a second edition in 1730), where he illustrated a very detailed refutation of Spinozism with an account of a dispute between Pythagoras and Anaximander. Then, in 1735, he published *Le Psychomètre ou Réflexions sur les Différents Caractères de l'Esprit par un Milord Anglais* in which he discussed the philosophical systems of Descartes, Newton, More, Spinoza, Malebranche, Locke, Bayle and Fontenelle. Actually these books were, in a certain way, a preparation for what Ramsay called the *Great Work*, that is to say his real and big project of writing a monumental work (in two parts) in which he tried to prove *more geometrico* the great principles of natural religion. The result of this project was *The Philosophical Principles of Natural and Revealed Religion* published posthumously between 1748 and 1749; unfortunately only in 1751 was this work reviewed in the *Monthly Review* and, notwithstanding the attention of Ramsay's friends (for example Francis Hutcheson and John Stevenson) and his widow, the book was not a great success and only few copies of it were sold. This is the main reason why Ramsay's ideas (with the most important accusation of Spinozism that Berkeley ever received) were forgotten for such a long time.

Ramsay had a profound aversion for Spinoza (he even affirmed that "Spinoza was not the 'God intoxicated' but the very worst of Atheist"[27]); but this polemic had deeper reasons, which were related to the real roots of his philosophy and, in a special way, they depended on Ramsay's relationship with Fénelon. He had convinced Ramsay that the Catholic religion was the true solution for his spiritual problems; but he was always very critical toward the Church of Rome, and always denounced its attempts to use faith for political purposes. He conceded to Catholic orthodoxy only the fact that it proposed an alternative to atheism. The book *Les Voyages de Cyrus* (VdC) does not end with the triumph of Christianity, even if the *prisca sapientia* is shown to be a fundamental educative instrument for the young Cyrus. We can conclude, therefore, that the VdC is a kind of allegory that represents Ramsay's brave search for religious truth.

---

[26] See G. D. Enderson, *Chevalier Ramsay* (Toronto and New York: T. Nelson S., 1952); M. L. Baldi, *Verisimile, non Vero. Filosofia e Politica in Andrew Michael Ramsay* (Milano: Franco Angeli, 2002); M. L. Baldi, "Andrew Michael Ramsay. Ciclicità e Progresso nell'Antica Teologia alle Soglie dell'Illuminismo," *Rivista di Storia della Filosofia* XLIV (1989): 443–476 and "Confutazione di Spinoza e Pirronismo. La Via al Senso Comune di A. M. Ramsay," *Rivista di Storia della Filosofia* 2 (1994): 215–261. See also D. P. Walker, *The Ancient Theology. Studies in Christian Platonism from the Fifteenth to the Eighteenth Century* (London: Duckworth, 1972): 231–239 and Cherel, *Un Aventurieur Religieux*, 1–64.

[27] Enderson, *Chevalier Ramsay*, 215.

The *Philosophical Principles*[28] (PP) is not unique; nor is it the first work in which the *Chevalier* accused Berkeley of Spinozism. In the first volume of the book (Chapter III: "*Of the Properties of Finite Beings*"), there is the *scholium* devoted to Berkeley's philosophy[29]; here the author discusses its similarities with Spinoza's *Ethica*. Ramsay's words are severe and inflexible:

> The learned Doctor Berkeley from a sincere and pious zeal against the absurd system of the materialists is the only modern author we know that has ventured to deny not only the real existence of bodies but even the possibility of their creation. [...] The negation of this principle tends necessarily to prove that God is the only agent in nature; and this opinion leads naturally to Spinosism because it induces us to believe that God is also the only substance in nature; as shall be fully unfolded in the next proposition.[30]

After the quotation of a long passage from DHP, in which Berkeley asserts that God eternally knows all the objects or, which is the same, that they have an eternal existence in His mind, Ramsay defines that as "the most refined Spinosism that ever was."[31] He concludes: "[...] this is the flower and the quintessence of Spinosism," even if "No doubt the great and the good Doctor did not perceive those fatal consequences of his scheme."[32]!

Before a rigorous discussion of every single point in Ramsay's line of reasoning, it is interesting to observe how his opinion was already present in the 1730 edition of VdC and how it is repeated, without corrections, in the second *Dialogue*.[33] In the VI book of VdC (1730), Ramsay studied the relationship between idealism, materialism and Spinozism through the exchange of ideas between two characters who defend different philosophical positions: Anaximander and Pythagoras. First of all, the *Chevalier* points to the authors who, according to him, unconsciously embraced Spinozism:

> La matière & l'étendue sont la même chose [*Descartes*, Ramsay's note]; or vous savez que l'étendue n'a ni couleur, ni odeur, ni saveur, j'ajoute qu'elle n'a ni bornes fixes ni parties distinctes, ni mouvements réels; toutes ces qualités ne sont que des idées ou des perceptions de l'âme [*le Docteur Berkeley*, Ramsay's note] causées par l'action de l'étendue immense qui se montre à nous successivement sous plusieurs formes différentes [*Le P. Malebranche*, Ramsay's note].[34]

Then, in a note to the dialogue, he explains the reasons for his judgement: "Voilà l'usage que les Spinozistes on fait du système de M. Descartes, du P. Malebranche & du Dr. Berkeley Anglais contre les intentions de ces trois Philosophes. [...] Le

---

[28] *The Philosophical Principles of Natural and Revealed Religion Unfolded in a Geometrical Order by the Chevalier Ramsay Author of The Travels of Cyrus* 2 vols. (Glasgow: Robert Foulis, 1748).

[29] Ramsay only refers to DHP without quoting TK.

[30] Ramsay, *The Philosophical Principles*, 243.

[31] Ibid., 244.

[32] Ibid., 245.

[33] We have to remember that the complete series of the *Dialogues* was never published. Now it is partially conserved in some manuscripts which belong to the Aix-en-Provence public library; only recently, M. L. Baldi published them in the book devoted to Ramsay that I quoted above.

[34] Ramsay, *Voyages de Cyrus*, ed. 1730, VI, 235–236. Here quoted from Baldi, *Verisimile, non Vero*, 356.

troisième tâche de prouver qu'il n'y a point des corps hors de nous & que tout est esprit [...]."[35]

Ramsay reflects upon the necessary distinction between immanent and emanating acts, thereby coming to an anti-pantheistic conclusion; this is that, in God, only what belongs to His essence is necessary, while the production of finite beings is a free and contingent act. Spinoza, in Ramsay's opinion, developed a philosophical system in which fatal necessity denies free will. Here, the idealist evolution of Spinozism is not yet clear and definitive, but it seems to give rise to the idea that the Spinozian heresy originates from the negation of the third substance, which is situated between the Infinite Spirit and created things. This conclusion will be definitive and clear in the PP:

> [...] par le Docteur Berkeley et les idéalistes que les figures, les divisions et les mouvements de la matière sont de pures sensations en nous comme les couleurs, les sons et les odeurs, et qu'il n'y a point de substance tierce entre Dieu et les esprits crées; [...] Il n'y a et il ne peut y avoir aucune substance tierce entre l'esprit infini et les esprits bornés, entre Dieu et les idées consubstantielles. C'est la même substance qui se présente à nous par conceptions partiales, qui est tout ensemble intelligente et etenduë et qui excite en nous par son action toutes les sensations [...].[36]

In the *Scholium*, devoted to this particular pantheistic evolution of idealism, Ramsay observes that man lives in a fallen state; in other words, he is alternately prey to pleasure and pain because he is joined with the material substance: "We shall show very soon that we are at present in a degenerate state where we are for wise reasons subjected to pain and pleasure by our physical union with material nature."

But, according to the idealistic perspective, if we erase matter (which generates pain, limitation and torment for limited beings), we will have to consider God as being responsible for sin and evil, since He is the only and infinite Cause: "But to suppose that God is the immediate author of all the impure imaginations, voluptuous sensations, and tormenting perceptions that expose the mind to corruption and depravation [...]."[37]

If we remove concreteness (that is to say, the material substratum) from objects, and if we consider them as a group of sensations of different kinds – Ramsay continues – nature will become a sort of phantom; it seems to be a collective illusion in which ideas substitute for concrete beings, forming a universe which is opposite to the cosmos described by the Holy Scriptures:

> Why then did God make this useless and arbitrary connection betwixt the ideas of sizes and figures and the sensations of pain and pleasure? All nature becomes a phantom, a mere delusion, a false appearance; there is nothing but ideas. This system therefore is equally opposite to God's eternal truth and wisdom, as also to his sanctity and goodness.[38]

---

[35] Ramsay, *Voyages de Cyrus*, ed. 1730, VI, 238. Here quoted from Baldi, *Verisimile, non Vero*, 357.

[36] Ramsay, *Extrait du 2^{d} Dial.*, here quoted from Baldi, *Verisimile, non Vero*, 442–443.

[37] Both quotations from Ramsay, *The Philosophical Principles*, 238.

[38] Ibid., 239.

Precisely at this point Ramsay introduces *Doctor* Berkeley who, notwithstanding the impious consequences of his immaterialist philosophy, is still considered sincerely religious by the author. In other words, Ramsay comprehends the apologetic reasons that inspire his attempt to fight against the materialists; but, at the same time, he denounces the absurd consequences that will result if we follow the Bishop of Cloyne's philosophical perspective until the end. According to Ramsay, Berkeley's argument against matter rests upon his considering the secondary qualities as equal to the primary ones; if we admit this, we will reach the (albeit erroneous) conclusion that matter does not exist. Berkeley confused ideas with objects, and perceptions with the true causes of them: "The Doctor's mistakes proceed from his confounding the ideas with the objects, and the perceptions produced with the cause producing. Simple ideas as we shall show are not objects which the mind perceives without itself but modes produced in it by the objects different from it."[39]

A closer analysis can show the core of Ramsay's criticism of Berkeley better: he is culpable of the elimination of any other cause but God; in this way, therefore, he created a system inspired by the principle later called "spiritual monism", which is also the fulcrum of Robert Clayton's accusation: "The Doctor does not pretend that it is finite spirit that excited in us all these perceptions but God alone."[40] In other words, if you deny that any other subjects (especially spirits) have the power to create something, you will be obliged to admit that the only true agent in nature is God.

In Ramsay's opinion, this is the pure essence of Spinozism:

> The negation of this principle tends necessarily to prove that God is the only agent in nature; and this opinion leads naturally to Spinosism because it induces us to believe that God is also the only substance in nature; as shall be fully unfolded in the next proposition. The ingenious and learned Doctor endeavours to prove in his third and last dialogue that God is the immediate cause and even the object of all our sensations; and that there is no third substance betwixt God and finite spirits which as we suppose acts upon us when we perceive extension, figure and motion.[41]

Ramsay refers to a long collection of passages taken from different places in DHP and joined together; from this, anyway, it is possible to clearly understand the target of the accusation, especially when the Irish philosopher describes God as an eternal and omnipotent Mind, which knows and comprehends ideas (that is to say beings, in Berkeleian terms) from the eternity.[42]

---

[39] Ibid., 241.

[40] Ibid., 241.

[41] Ibid., 243.

[42] In the *collage* of quotations from Berkeley's DHP here proposed by Ramsay we can read: "The eternal, omnipresent mind, says he, which knows and comprehends all things, exhibits them to our view in such a manner and according to such laws, as he himself has ordained, and are by us called the laws of nature. All things exist in the divine mind from eternity. When things are said to begin or end their existence, we do not mean this with regard to God, but the creatures. All objects are eternally known by God, or which is the same thing have an eternal existence in his mind: but when things before imperceptible to created spirits are by a decree of God made perceptible, then

Geneviève Brykman carefully studied this charge, which seems effectively to discover a delicate and crucial point of Berkeley's philosophy. She argues that the rapport between Berkeley and Spinoza can be studied starting from two points of view: first of all, we have to remember that Berkeley had surely read Spinoza's *Opera Posthuma* (as we can see from *Philosophical Commentaries,* PC). Secondly, if we analyze the quotations concerning the Amsterdam philosopher taken from all Berkeley's works, we will realize that they constantly refer to a denunciation of Spinozism. In TK, DHP and *Alciphron,* Spinoza's name only appears in the brief list of atheists detected; but, from what we can see from PC, the philosophy of Spinoza seems to have played an important role in Berkeley's early years; Brykman even thinks that the elaboration of idealism comes directly from the reading of Spinoza's *Ethica* and *Letters.*[43] I completely agree with Brykman also when she claims that Berkeley had known Spinoza's philosophical system through the reading of the article *Spinoza* in Pierre Bayle's *Dictionary.*[44]

If we join together these considerations concerning the doctrine of the unique substance with the principle of Ramsay's accusation, we will clearly understand how the author of PP was, at the same time, defending his *personal* doctrinal position. In his first books, indeed, he always tried to demonstrate the necessary existence of a third substance which is intermediate between God and finite beings; consequently, he insisted on the impossibility of the existence of only one and immediate cause in nature: "[...] we have already shown that the infinite mind cannot be the

they are said to begin a real existence with respect to created minds. By creation therefore, nothing else can be understood but that the several parts of the world already existent from all eternity in the divine mind, become gradually perceptible to finite spirits endued with proper faculties." (Ramsay, *The Philosophical Principles,* 243).

[43] G. Brykman, "Berkeley, Lecteur et Critique de Spinoza," *Recherches sur le XVIIème Siècle, Éditions du Centre National de la Recherche Scientifique, Cahiers de l'Equipe de Recherche,* n. 75 (1978): 175. Of the same author, (whose studies about the relationship Berkeley-Spinoza are still an unavoidable point of reference) see also: G. Brykman, dir. "Berkeley et le Cartésianisme," *Le Temps Philosophiques,* Université Paris X, Nanterre (1997); "Berkeley et l'Intérieur Absolu des Choses," *Revue Philosophique de la France et de l' Étranger,* 4 (1980): 421–425; "Berkeley on "Archetype", in *Essays on the Philosophy of George Berkeley,* ed. E. Sosa, (Dordrecht: D. Reidel, 1987): 103–112; *Berkeley. Philosophie et Apologétique* (Paris: J. Vrin, 1984); "Berkeley: sa Lecture de Malebranche à travers le *Dictionnaire* de Bayle," *Revue Internationale de Philosophie,* n. 4 (1975): 496–514; "Du Commencement Introuvable de l'Immatérialisme," *Les Études Philosophiques,* n 1 (1980): 385–397; "Le Model Visuèl de la Conaissance chez Berkeley," *Revue Philosophique de la France et de l' Étranger,* n. 4 (1983): 427–441; "Microscopes and Philosophical Method in Berkeley," in *Berkeley. Critical and Interpretive Essays,* ed. C. Turbayne (Minneapolis: University of Minnesota Press, 1982): 69–82.

[44] As to the relation between Bayle and the English cultural world, see L. P. Courtines, *Bayle's relations with England and the English* (New York: Columbia University Press, 1938). From the *Dictionnaire* it is possible to grasp the complete knowledge that Bayle had of England and a very interesting list of the English authors and their works. One ought to delve further into the complex relation between Berkeley and Bayle, but the limits of this article do not allow it. Cf. G. Brykman, "Berkeley: sa Lecture de Malebranche à travers le Dictionnaire de Bayle," *Revue Internationale de Philosophie,* n. 4 (1975): 496–514.

immediate cause of these ideas, sensations and modifications; and therefore there must be a third substance betwixt God and human soul that really acts upon us to produce in us different sensations, according to our different organs, states and situations."

As a result, it is impossible to renounce matter because it is the cause of our perceptions; the fact that there is a difference among sensations (due to the variation of organs and distances), is not a good reason to refuse to admit the existence of objects. The *Chevalier*'s words look like an admonition toward Berkeley: "But still there will be real matter, the cause and the object of their perceptions."[45]

In any case, according to Ramsay, materialism and idealism are two roads, surely opposite, which lead to the same result, that is to say the affirmation of the theoretical core of Spinozistic theory: "Otherwise we must fall naturally into Idealism or Materialism which by different roads lead at last to the Spinosian impiety, which asserts that nature is composed of one only substance of which all beings are only modalities, hypostases or consubstantial forms."[46]

It is worth delving further into this point, which seems to be one of the most interesting of the book. Ramsay considers materialism and idealism as the two main lines of the reflection that arrives at Spinozism; for this reason he tries to confute them, even if, at the same time, he admits that idealism is based on a theoretical analysis, which is very difficult to deny. The following passage shows how Ramsay deals with this question from the first pages of PP:

> Tho' these two sentiments appear contradictory yet they both equally tend to Spinosism. For if once we suppose that there is and can be but one sort of substances, it is easy from thence to pass to belief that there is but one only substance in nature, which is the essence of spinosian scheme. It is of great consequence therefore to confute both these systems, and to prove, first, against the defenders of each, that God can create two sorts of substances material and immaterial, whose essential properties are not only different but contradictory and incompatible. We shall show in the next place against the Materialist that motion is not essential to matter; and that intelligence and extension cannot be properties of the same substance. We shall endeavour to prove in the last place against the Idealist, that God cannot be the immediate author of all the ideas, sensations and perceptions we have of matter [...].[47]

Ramsay is inclined to admit that Malebranche opened the road to idealism; nevertheless he did not commit the same error as Berkeley did: denying the existence of matter had become "the most embarrassing paradox of modern philosophy."[48] The *Chevalier* proved that Berkeley was *the most refined Spinozist* because he used the same fundamental principles of that theory:

> This is the most refined Spinosism that ever was; for here the Doctor adopts three of the greatest fundamental principles of that impious scheme, (I) The identity of ideas and their objects; (II) The coeternity of these ideas in the divine mind, and consequently their

[45] Both quotations are from Ramsay, *The Philosophical Principles*, 241.

[46] Ibid., 247.

[47] Ibid., 190.

[48] Ibid., 384.

> consubstantiality with God; (III) The uselessness of creation. We have already demonstrated (1) that the archetypal ideas of things finite are quite different from the objects themselves; (2) that is absolutely false that all finite ideas have an eternal and necessary existence in the divine mind; (3) that creation is not only possible, but absolutely necessary to explain the phenomena of nature produced.[49]

Even if Berkeley wanted to defend the doctrine of Creation, he asserted a variation of it, which, deprived of the function that historically belonged to matter, is absolutely different from what was traditional. As a result – Ramsay concluded – the Universe of this philosophy evokes the cosmological concept of emanation:

> Thus nature produced [...] must be looked upon as an emanation of the divine substance, which by immanent action produces all in itself [...] thus all ideas of creation are destroy'd; and what we mean thereby is only that the coeternal consubstantial forms or ideas of the divine mind being living, self-conscious, intelligent hypostases or personalities are affected with different sensations, perceptions and modifications produced by the immanent action of the absolute infinite upon itself; this is the flower and quintessence of Spinosism.[50]

We have to consider now the last argument that Ramsay used against Berkeley. After examining how Berkeley rejected Malebranche's occasional causes, Ramsay observed that in the books of the Irish philosopher matter "cannot be an occasion so it cannot be a simple instrument of the divine operation." Since God is considered the unique and immediate cause of everything that happens in nature, He should become responsible even for sin and evil in the world; as a result, also the creatures' moral responsibility is denied, because they cannot decide differently from what has already been decided: "This argument [...] destroys the activity of second causes and supposes that God is the immediate author not only of all the physical actions, but of all the moral determinations of created beings. Thus liberty is destroyed and God made the sole cause of all the sins and blasphemies of finite spirits." It is obvious – Ramsay continues – that finite beings have to be joined with their cause through a certain relation of dependence; but this does not mean that: "all actions have such an immediate and absolute dependence on the divine action."[51]

These words seem to reflect a theme which Ramsay closely studied from the start of his first books; that is to say, the battle against predestination. Without the divine Providence, in effect, religion and even the very concept of God could lose their meaning, as "Eternal Providence desires wills and employs continually all the means necessary to lead intelligent creatures to their ultimate and supreme happiness."[52] Ramsay's ambitious project consists of unifying liberty (used by the creatures in order to self-determine their destiny) with the ineluctable necessity of actuating the divine decision. Notwithstanding Berkeley's apologetic declarations, his consequences appear in fact, in Ramsay's eyes, to be in contrast with the true

[49] Ibid., 244.

[50] Ibid., 244–245.

[51] Ibid., 242.

[52] Ibid., 181.

principles of the traditional faith. And, in effect, where is the necessary distinction between the divine and natural dimensions? This dissimilarity is also essential in order to legitimate men's freedom and justify their efforts to behave properly, while, in the Berkeleian system, the omni-pervasive action of God seems to fill all the disparities. We are faced with the extreme consequences of idealism: it wanted to make a contrast with materialism, but in reality it appears only as its complementary opposite. In both cases, indeed, the acknowledgment of a unique substance (whether material or spiritual is only a detail), leads to a denial of any possibility of choice for creatures.

Ramsay was very severe in his accusation but, in any case, he always recognized Berkeley's pious purposes: "The learned Doctor Berkeley from a sincere and pious zeal against the absurd system of the materialist [...]."[53] His zeal was so sincere and fervid that it prevented him from seeing the limits which he was unintentionally surpassing: "No doubt the great and good Doctor did not perceive those fatal consequences of his scheme."[54] In other words, the purposes of Berkeleian metaphysics were certainly to fight against the tendency to amplify the importance of geometry and the role of matter; but the methods and, above all, the results of this attempt inevitably favoured Spinoza's system.

12.4 Robert Clayton

Three years after Ramsay's accusation, another theoretical attack on Berkeley's philosophy came from *An Essay on Spirit* by Robert Clayton, Bishop of Clogher.[55] Clayton had been Berkeley's friend since their time at Trinity College, and in 1737 he also tried to introduce him to the Irish House of Lords; they were responsible for two dioceses very close to each other but, above all, they were companions in the Bermuda project. This last biographical element helps us to better understand their relationship which was sometimes complex and difficult. We know that the Berkeleian project of building a Bermuda College in the American colony of Rhode Island (whose head had to be Clayton), also failed owing to Clayton's abandonment, having been elected Bishop of Killala. Many hypotheses have been made to explain this sudden alteration of their friendship. David Berman suggests that it happened around 1752, when both Berkeley and Clayton exchanged negative judgments concerning their respective works; moreover, it is important to remember

---

[53] Ibid., 239.

[54] Ibid., 245.

[55] Robert Clayton, *An Essay on Spirit, wherein the Doctrine of the Trinity is considered in the Light of Nature and Reason...* (London, 1751). See the works of D. Berman, *George Berkeley. Idealism and the Man* (Oxford: Clarendon Press, 1994), and "Berkeley, Clayton and *An Essay on Spirit*," *Journal of the History of Ideas*, XXXII (1971): 367–378. The author is convinced that commentators always undervalued not only the Irish philosophical tradition, but also the influence that it had on Berkeley (Berman, *Idealism and the Man*, 9).

that Clayton addressed his criticism principally to *Siris*, one the most controversial of Berkeley's books.

The *Essay* came out anonymously and only in a few copies (about 40), all sent to the most influential Irish bishops. Another five editions followed, but only in 1752 was the book finally attributed to Clayton. There are only two places in which the author quotes Berkeley, although it is possible to follow the argumentation in the polemic. The first mention is in the *incipit* that comes after the long *Dedication*; it linked Berkeley and Malebranche in the accusation of embracing and cultivating Spinoza's heresy:

> The opinion of Spinoza was that there is no other Substance in Nature but God; that modes cannot subsist or be conceived without a Substance; that there is nothing in Nature but modes and substances: and that therefore every thing must be conceived as subsisting in God. Which with some few alterations hath been embraced and cultivated by P. Malebranche and Bishop Berkeley.[56]

With the exception of a note in which Clayton quotes the passages from *Ethica* only mentioned in the text, there are not many other analyses or references to both authors' philosophies; only in the second and third paragraphs, in which he explains the principle and the consequences of the theory of *Self-Existing Being*, does there seem to be an indirect allusion to the Berkeleian doctrine. Without quoting the Bishop's name, he alludes to the "Substance" and to its "action of exciting ideas to our minds": all that in a context in which he is talking about Spinoza's philosophy: "[...] Existent Being can raise Ideas in our Minds, or, which is the same thing, can become knowable by us. Every Existence or Being I therefore call a Substance; the Manner in which it makes an Impression on our minds I call a Mode; and the effect or Impression which is thereby made upon the Mind I call an Idea."[57]

The bibliographical source from which Clayton took his critical consideration[58] is *Siris*: "Which plainly proves that Resistance is something more than inability, or a Want of Power or a negation of Spirit as the Author of *Siris* asserts it only be."[59] That is to say: matter, denied by Berkeley, exists, has qualities, and is not only the negation of spirit.[60] It is worthwhile, now, to study Clayton's doctrine in its essential points, in order to better understand the theoretical reasons behind his criticism.

At the top of his metaphysical system there is God, the Supreme Author of the Universe, the sole being who does not suffer the limitation imposed by the conjunction with matter:

> And as the Almighty God is the only supreme, infinite, unlimited, Being in the Universe; so is he, probably, the only unembodied Spirit that exists: that is, the only Spirit which is

[56] Clayton, *An Essay on Spirit*, 1–2.

[57] Ibid., 2.

[58] Berman suspected that Clayton appropriated some of the ideas of *Siris* (as, in effect, Clayton admitted in a *pamphlet* published in 1754). Cf. Berman, *Idealism and the Man*, 185–186.

[59] Clayton, *An Essay on Spirit*, 10.

[60] Cf. Yolton, *Thinking Matter*, 97–98.

> not limited, clogged, and settered, with some kind, or Degree, of inactive Matter, which may serve to give a Form and Shape or Boundary to its Spiritual Nature.[61]

By just considering this point, we can understand the fundamental value that Clayton gives to matter, the passive substratum for the spirit's action; the philosophical system that results has a clearly dualistic foundation, perhaps inspired by Descartes:

> And as my own consciousness convinces me of my own Existence so does the same Faculty convince me that this Existence of mine is composed of two very different Kinds of Existence, that is, of a thinking, active, powerful, Existence; and a dull, heavy, inactive, Existence. One of which [...] call the Spiritual Existence, Subsistence or Substance; and the other, viz., the inactive, we will call the material or bodily Existence; and sometimes, for brevity, we will call one Spirit and the other Matter or Body.[62]

Matter is not able to make any movement because activity only belongs to the spirit: "[...] that kind of Existence, which we call Matter, is incapable of producing any kind of Motion either voluntary or involuntary"[63]; but the most important capacity of its nature is of *infinitely receiving in itself* the action of Spirit, like His physical substratum.[64] Therefore, it is evident that Clayton cannot accept Berkeley's point of view; if it had won out, the entire philosophical system of the *Essay* would have been destroyed from its very foundation.

In Clayton's theoretical universe, there is a place also for *Logos*, that is to say *God's Word* (also called *Angel*): "And as this Angel [...] is called the Word of God because God employeth him to carry his Word; so he is also, by the same figure or rhetoric called the Wisdom of God."[65] It is the first among a myriad of spiritual agents which realise God's Will; although they are very powerful, they do not have their being per se, so they totally depend on Him:

> And as God may communicate what Proportions he pleases of his Attributes to the different Gradations of created Beings, with which he hath been pleased to fill in the Universe. [...] And hence it is, that human beings may be surrounded with myriads of Spiritual Agents without ever being sensible thereof; unless those superior beings are pleased to assume such Forms and condescend to furnish themselves with such Qualifications, as are capable of making an Impression on the human Spirit from within or the human senses from without.[66]

---

[61] Clayton, *An Essay on Spirit,* 27–28. At the end of the book, Clayton discussed the concept of God in the old way; he tried to demonstrate that Plato's and Pythagoras' philosophies are very close to the old Egyptian doctrines which often defined divinity as a Mind, Reason, Wisdom. It is very interesting to observe that sometimes we can also find these definitions in *Siris.*

[62] Ibid., 6.

[63] Ibid., 8.

[64] "Clayton accepts the independence of mind and body. [...] Therefore Clayton believes every piece of matter has united to it an individual spirit, which governs and effects its movements. His theory might be described as a pluralistic version of occasionalism." Berman, *Idealism and the Man,* 182.

[65] Clayton, *An Essay on Spirit,* 50.

[66] Ibid., 31–32.

Therefore, the focal point of Robert Clayton's charge to Berkeley seems to concern the *monist* tendency[67] of that philosophy, in which, more than renouncing matter, all reality is pervaded and directed by God:

> [...] we cannot but think that God should chuse [sic] to govern this Universe by a gradual Subordination of Beings, one superior to another, rather than to be the sole Director or Governor of every the most minute Affair [...] But because it seems more consistent with the divine Goodness and Wisdom to employ the various Works of his Hands, in the Exercise of those Powers and Faculties with which he hath endowed them; rather than personally and immediately to interpose in the Conduct of those Transactions for which he hath created Numbers of Beings furnished with Abilities sufficient to perform them.[68]

Even if other commentators had individuated this same perspective, Clayton's position seems more functional to the defence of his proper doctrine; if he had accepted the Berkeleian philosophy, he would not only have dangerously approached Spinoza's philosophy, but also renounced the principle of plurality, the true basis of his system.[69]

12.5 Conclusion

The last phase in the reception of his works was particularly satisfying for Berkeley; it was characterized by epistolary exchanges and clever critical observations – not distorted, at least, by the usual satirical comments concerning immaterialism. This period reached his highest point with the publication, in 1754, of Samuel Johnson's *Elementa Philosophica.*

In 1751, Henry Home published an important book for the "fortune" of the Irish philosopher: the *Essays on the Principles of Morality and Natural Religion.*[70] The essay dedicated to Berkeley is the third: *Of the Authority of Our Senses*: the author directly and resolutely remarks on the effects of the denial of a material substratum. If we accept the principles of Berkeley's philosophy, we will reach two different conclusions that are both very far from the traditional religious doctrine: first of all, man becomes the unique being in the world; and, secondly, one of the most important demonstrations of God's existence is deprived of its fundamental premise.

---

[67] Berman emphasises that monism, as the main peculiarity of the Berkeleian doctrine, had been highlighted also in an article published in the *Literary Journal* of Dublin in 1745; in that text the author (whose initials D. G. S. could indicate *Dean Gervais*, one of the most important of Berkeley's friends and correspondents) quoted *Siris*, above all the paragraph n. 239 in which he said that all the reality depends on the immediate action of an Incorporeal Agent who moves and disposes of all the things according to His rules.

[68] Clayton, *An Essay on Spirit*, 85.

[69] D. Berman, *Idealism and the Man*, 183: "If the monism, in this sense, of Malebranche and Berkeley is right, then clearly Clayton's pluralism is wrong."

[70] The text, published in Edinburgh (1751) by Kineaid and Donaldson is quoted in its entirety in D. Berman (ed.), *Eighteenth-Century Responses*, 2 vols. (New York & London: Garland, 1989): vol. II, 237–269.

The interpretative model, proposed between 1718 and 1751, is now only a minor and nearly forgotten aspect in the traditional Berkeleian historiography. Contrary to the established icon of Berkeley as an apologist, in four specific cases (that is to say in the review of *Acta Eruditorum*, and in the works of Baxter, Ramsay and Clayton) he was openly accused of adhering to Spinozism and/or of gliding toward atheistical positions. Apart from the reviewer of *Acta Eruditorum* (who merely observed that *Ethica* could easily have been counted among the Berkeleian sources), the others provided their accusations with relevant reasons.

The first argument which they discussed was the refutation of material substance (the same that the public of that period considered so strange as to be incredible). Beyond the natural and obvious astonishment (which comes from thinking of the world without matter), the first commentators noticed that, in this way, God remains *the unique and infinite substance really existent*, in a philosophy dangerously close to Spinoza's system. This comment by G. Brykman perfectly shows the meaning of this parallel: "Si donc, en tant qu'homme d'Eglise, Berkeley a l'intention de combattre le Spinosisme, il avait, en tant que philosophe, l'intention de agir mieux que Spinoza."[71] Even if Berkeley thus intended to cut a fundamental argument in favour of the materialists, his attempt to mitigate the differences between the divine cause and natural effects (*spiritual monism*) was stigmatised first by the reviewers of the *Journal Litéraire* and then by Tournemine.

The majority of these authors indicated another consequence that comes from the rejection of matter: Baxter, Ramsay, Clayton and Home accused Berkeley of reducing the universe to a *collection of ideas* (instead of *concrete beings*); in this way he also transformed reality into an unstable illusion, deprived of its authenticity. Instead, the author of TK had tried many times to explain that he wanted to eliminate *only matter*, because it is a useless and contradictory substance; he absolutely did not want to deny the existence of sensible objects, as they constitute the whole *corpus* of natural phenomena. Evidently, in spite of his efforts, Berkeley did not succeed in convincing his readers that with his doctrine the Universe still conserves its traditional features of reality and authenticity.

The Jesuits of *Mémoires de Trévoux* and Ramsay observed another ambiguous consequence concerning the Berkeleian theory of the unique and infinite First Cause: confronting the doctrine of Creation in the version of Holy Scriptures with the Berkeleian description, they saw a deep and worrying disparity. This question is actually linked to the theory concerning ideas in God's mind (*archetypes*) and ideas excited in men's minds (*ectypes*). Even if Berkeley was not so clear in this respect, nevertheless a lucid problematic core comes out from it: (1) if (as Ramsay showed) all the beings in *rerum natura* are ideas, (2) if they share an eternal existence in God's Mind, and (3) if the only difference between divine and human ideas is their order but not their essence, then the Universe will appear more as the direct manifestation of the Divine Cause than as an effect of His Will.

[71] Brykman, *Berkeley, Lecteur et Critique de Spinoza*, 177.

Chapter 13
The Animal According to Berkeley

Sébastien Charles

Few interpreters of Berkeley's texts have taken an interest in the status of the animal in his work, which is easily explained by the fact that Berkeley himself hardly seemed concerned by it. Yet Berkeley's conception of the animal is not without its difficulties, which even some of his eighteenth century readers were already to notice. Andrew Baxter, for instance, in his *Enquiry into the Nature of the Human Soul* of 1733, wonders whether Berkeley's immaterialism must not necessarily lead to a denial of the capacity of any animal to perceive exterior objects, since perception supposes a reflective activity of the mind and Berkeley seems to deny any such activity to animals.[1] Meanwhile, Samuel Johnson, thinking that Berkeley does indeed recognize the capacity of animals to perceive and feel exterior objects, asks him in a letter of 5 February 1730 whether his conception of the animal does not lead to granting animals souls as immortal as those of humans.[2] It seems to me that such difficulties of comprehension were caused in part by the fact that Berkeley himself never explicitly dealt with the question of the animal in his work; it may also be the case that his equivocal use of the terms "mind" and "soul" in his writings would lead some readers to wonder about the difference between animals and men. Above all, however, such

[1] Andrew Baxter, *An Enquiry into the Nature of the Human Soul* (London: James Bettenham, 1733), 313. At this rate we must say, that brutes have no objects of their sensations, since sensations cannot be objects to themselves; for they make no reflex acts of the mind, and there are no material objects from without, according to this scheme.

[2] Letter of 5 Feb. 1730 from Johnson to Berkeley, in Luce-Jessop, *The Works of George Berkeley, Bishop of Cloyne* (London: Nelson, 1948–1954), II, 289–290.

S. Charles (✉)
Université de Sherbrooke, Département de philosophie et d'éthique appliquée,
J1K 2R1, Sherbrooke, Québec, Canada
e-mail: Sebastien.Charles@Usherbrooke.ca

S. Parigi (ed.), *George Berkeley: Religion and Science in the Age of Enlightenment,* International Archives of the History of Ideas/Archives internationales d'histoire des idées 201,
DOI 10.1007/978-90-481-9243-4_13,

difficulties may have been due to the influence, at the time when Berkeley's texts were first being read, of the Cartesian animal-as-machine thesis.[3]

Looking at the terms in which Descartes himself had framed the question of the status of the animal helps us clarify Berkeley's own position on it. As is well known, Descartes rejects the Aristotelian distinction between *anima*, the soul's sensitive faculty, and *animus*, the soul's intellectual faculty – a distinction which had allowed Aristotle to attribute souls to animals, yet distinguish the faculties of such souls from those of humans. Descartes rejects this distinction in favour of a radical dualistic logic: either we do have a soul, and every characteristic that belongs to it, or we do not. Since perception is a faculty of the soul, says Descartes, we must conclude that beings without souls do not really perceive, and thus experience no pleasure or pain, responding only instinctively to any exterior stimuli, much as an alarm-clock will ring at a precise time not because it perceives that such is the right time to ring but because it has been programmed to do so. To this effect, in a letter of 1637 to Plempius, Descartes writes "my view is that animals do not see as we do when we are aware that we see, but only as we do when our mind is elsewhere" – which is to say, as when we see something without paying attention to it because our mind is at that moment distracted by something else.[4] Animals do perceive, then, but not consciously.

The primary advantage of giving animals the status of machines is theological, in its definitively resolving the difficulty of the question of an immortality of the animal soul. Descartes's proof against the immortality of their souls is the fact that animals do not talk – even if they do seem to exchange information amongst themselves (or with men) by displaying what appear to be desires or fears – and that, since they do not talk, by inference they do not think; and if they do not think then they have no soul, since thinking is an essential faculty of the soul. Bernard Baertschi quite rightly calls this the *principle of equipotency of souls* – the principle that "every being that has a soul also has all its faculties, language included, and [that] any being who does not talk, even if he expresses things, is *ipso facto* a machine."[5] From the principle that souls do not differ by degrees it follows that if a being does possess a soul, its soul is necessarily immortal, a position which has the undeniable advantage, as Descartes would remind the Marquis of Newcastle in a letter of 23 November 1646, of not attributing immortality to oysters or sponges.[6]

---

[3] The controversy would involve not only Descartes, but many of the greatest names in philosophy of the later seventeenth and early eighteenth centuries, including Arnauld and Nicole, Cordemoy, Pascal, Rohault, Malebranche, Leibniz, Le Grand, Dilly, Lamy, Bayle, Régis, Fénelon, Locke, Sergeant, and others.

[4] Letter of 3 Oct. 1637 from Descartes to Plempius for Fromondus, in *The Philosophical Writings of Descartes, Vol. III* (Cambridge: Cambridge University Press, 1991), 61–62.

[5] Bernard Baertschi, *Conscience et réalité. Études sur la philosophie française au XVIII*[e] *siècle* (Geneva: Droz, 2005), 21. See also from the same author: *Les rapports de l'âme et du corps. Descartes, Diderot, Maine de Biran* (Freiburg: Éditions Universitaires, 1992), 66.

[6] Letter of 23 Nov. 1646 from Descartes to the Marquess of Newcastle, in *Philosophical Writings, Vol. III*, 302–304.

It is because his readers have this mechanistic, Cartesian conception of the animal in mind that they do not immediately grasp Berkeley's thought. I take as a proof of it the aforementioned queries of Samuel Johnson in his correspondence with Berkeley:

> I think I once heard you allow a principle of perception and spontaneous motion in beasts. Now if their *esse* as well as ours consists in perceiving, upon what is the natural immortality of our souls founded that will not equally conclude in favour of them? I mention this last consideration because I am at a loss to understand how you state the argument for the soul's natural immortality.[7]

The problem posed by Johnson is simple: since immaterialism presents itself as a dualism between perceiving subjects and perceived objects, to confer perception upon animals must be to assume that animals do have minds, and therefore souls, and indeed souls presumably as immortal as our own, as the Cartesian principle of the equipotency of souls demands. But is that what Berkeley thinks? Here we must withdraw from the dualistic logic and animal automatism of Descartes and see fully the difference between the philosophies of Descartes and Berkeley as they concern the question of the status of the animal.

Let us take as our point of departure the question of animal perception. As we have seen, Samuel Johnson thinks he once heard Berkeley attribute both a principle of perception and a form of spontaneous motion to animals; and in this he is correct. Berkeley's work is replete with analogies between animal and human perception, leaving no doubt that he believes in the capacity of animals to perceive and to interact with the world around them by means of their own senses. For Berkeley, the general functioning of the senses is the same throughout the order of living beings and is how each being preserves itself, as Hylas must concede to Philonous:

> PHILONOUS. – Answer me, Hylas. Think you the senses were bestowed upon all animals for their preservation and well-being in life? Or were they given to men alone for this end?
>
> HYLAS. – I make no question but they have the same use in all other animals.
>
> PHILONOUS. – If so, is it not necessary they should be enabled by them to perceive their own limbs, and those bodies which are capable of harming them?
>
> HYLAS. – Certainly.[8]

From this we can also conclude that the perception of pleasure and pain are common throughout the animal kingdom, which Berkeley himself effectively claims in Section 59 of the *New Theory of Vision* by insisting on the preventative role of vision. Every animal is sensitive to modifications of his environment and to what advantages or disadvantages such changes may pose to his mind no less than to his body, since his mind is what receives the signals of pleasure or pain so indispensable to his well-being. From this it can be argued that animals have a form of experience that lets them take best possible advantage of their natural environment.

---

[7] Letter of 5 Feb. 1730 from Johnson to Berkeley, in *Works*, II, 289–290.

[8] Berkeley, *Three Dialogues between Hylas and Philonous*, in *Works*, II, 188.

> For this end the visive sense seems to have been bestowed on animals, to wit, that by the perception of visible ideas (which themselves are not capable of affecting or any wise altering the frame of their bodies) they may be able to foresee (from the experience they have had [of] what tangible ideas are connected with such and such visible ideas) the damage or benefit which is like to ensue, upon the application of their own bodies to this or that body which is at a distance.[9]

But does the attribution of perception to animals not pose a serious problem within the framework of an immaterialism for which perception supposes a perceiving mind – which is also to say a soul, if we hold to the definition of the *Principles* stating that "this perceiving, active being is what I call *mind, spirit, soul* or *myself* (by which words I do not denote any one of my ideas, but a thing entirely distinct from them, wherein they exist, or, which is the same thing, whereby they are perceived)"?[10] The same would also seem to go for pain, since only the mind can feel it: "Nobody will pretend that real pain either is, or can possibly be, in an unperceiving thing or without the mind, any more than its idea," says Berkeley elsewhere.[11] These questions only lose their aporetical character if we withdraw from the Cartesian framework and accept the fact, as Berkeley seems to do, that the difference between men and animals is, above all, a question of degree and not one of nature, and that countless varieties of minds exists, including minds inferior and superior in capacity to those of humans. We must then legitimately recognize that animals could possess both minds and souls, even if it remains to be known what meanings should be accorded such terms, or what ontological difference Berkeley may conceive to exist between men and animals.

From the *Philosophical Commentaries* to the *Siris*, one idea running through Berkeley's work is that of a chain of beings of such imperceptible gradations from one rank to the next that Berkeley sometimes brings childhood near to animality. This implies that at the level of creatures differences are of degree and not of nature: "If you take away abstraction, how do men differ from beasts? I answer: by shape. By language rather by degrees of more and less."[12] But if the difference between man and animal is merely one of degree, as Berkeley supposes, then in spite of everything we must recognize a fundamental distinction between the ranks of animals and that of men, and their being distanced enough that the most intelligent of animals could never equal the stupidest of men: "We imagine a great difference and distance in respect of knowledge, power, etc., between a man and a worm. The like distance between man and God may be imagined or infinitely greater."[13]

In reality, the qualitative differences between men and beasts are relatively numerous and are mainly differences between the faculties of their respective souls or minds. Inspired less by early modern dualism than by the dualism of antiquity which distinguishes a rational soul from a sensitive soul and allows for the attribution

[9] Berkeley, *Essay Concerning Vision*, § 59, in *Works*, I, 193.

[10] Berkeley, *Principles of Human Knowledge*, § 2, in *Works*, II, 41–42.

[11] Ibid., § 41, 58.

[12] Berkeley, *Philosophical Commentaries*, in *Works*, I, A594, 74.

[13] Ibid., A640, 78.

of sensation to the entirety of the living order, Berkeley attributes as much sensation to animals as he does imagination, explicitly distancing himself from the Cartesian way on this point:

> [The Pythagoreans and Platonists] accurately considered the differences of intellect, rational soul, and sensitive soul, with their distinct acts of intellection, reasoning and sensation, points wherein the Cartesians and their followers, who consider sensation as a mode of thinking, seem to have failed.[14]

Moreover, it is sensation that carries the day for animals, which is why they seem entirely guided by what affects them and never to attain the reflective level they would need in order to wonder about what they perceive. While animals, being perceivers, must have ideas, their knowledge can go no further than whatever the data their senses transmit, which is why complex ideas like *unity* or *existence* seem entirely to escape them.[15] So too do their imaginations seem just as relatively limited; Berkeley will go so far as to ask whether an essential difference between men and beasts might not lie in the capacity of men to elaborate upon sensible ideas: "Question: whether composition of ideas be not that faculty which chiefly serves to discriminate us from brutes. I question whether a brute does or can imagine a blue horse or chimera."[16]

Omnipresent sensation, weak and limited imagination: here is a preliminary sketch of the animal according to Berkeley. Should animals also be granted a share in reason, if only a limited share, and one that varies from one species to another? Another claim of Berkeley's is that "to perceive is one thing; to judge is another" – from which it might be argued that animals have no ability to reason.[17] With the principle of ontological continuity proper to living beings in mind, however, might we not also come to the opposite conclusion? The *Siris* seems headed in that direction, suggesting a steady continuity between the sensitive and the intelligible: "So as in the rational animal there is still somewhat intellectual, again in the sensitive there is somewhat rational, and in the vegetal somewhat sensitive."[18] Here Berkeley merely puts forward the doctrine of Plato without necessarily endorsing it. If Berkeley's definitions of the soul and of the mind as given in the *Principles* are adhered to, however, animals would have to be accorded some form of thought and reflection. In fact, does Berkeley not on numerous occasions present the perception of ideas as being inextricable from both thinking and acting?[19] Does he not insist numerous times that the mind is so simple that any distinction among its faculties

---

[14] Berkeley, *Siris*, § 266, in *Works*, V, 125.

[15] See Berkeley, *Philosophical Commentaries*, A746, in *Works*, I, 91. "Will any man say that Brutes have ye ideas, unity and existence? I believe not. Yet if they are suggested by all the ways of sensations, tis strange they should want them."

[16] Ibid., A753, 92.

[17] Berkeley, *The Theory of Vision Vindicated and Explained*, in *Works*, I, 265.

[18] Berkeley, *Siris*, § 275, in *Works*, V, 129.

[19] Berkeley, *Principles*, § 27, in *Works*, II, 52. "A spirit is one simple, undivided, active being: as it perceives ideas, it is called the *understanding*, and as it produces or otherwise operates about them, it is called the *will*."

can only ever be a rational contrivance, and never real?[20] Does he not recall that no middle ground is to be had between the mind and its ideas, with the consequence that animals must be placed on the spiritual side?[21] Finally, does he not submit that the nature of the mind, or soul, is to always think?[22]

It is true that the above passages from the *Principles* seem concerned above all with the human mind, and that any question of the existence of animal thought is mentioned only in passing, in the *Dialogues*, when Philonous, stating that animals *should think* that the shape and extension they touch belong to exterior objects, is cautioned by Hylas that one does not know whether animals think or not.[23] If only the Cartesian criterion of language could be taken as what, when present, reveals thought, Berkeley would then side with Descartes and in the absence of any evidence of significant language among animals deny them any thought. This is made clear notably in Sections 16 and 17 of the Introduction to the *Principles* in its original manuscript, and in virtually identical form in Section 11 of the Introduction as published in 1710, where Berkeley takes issue with Locke's abstract general ideas that would supposedly differentiate animals from men.[24] In rejecting even the possibility of abstraction Berkeley cannot let their having abstract ideas be what makes humans specifically different. Only the power to subsume many particular ideas under a single general term seems to be the genuine privilege of human beings, and seems to testify to their being the only ones who truly reason. This is confirmed by Euphranor in a passage from *Alciphron*:

> All signs are not language: not even all significant sounds, such as the natural cries of the animal, or the inarticulate sounds and interjections of men. It is the articulation, combination, variety, copiousness, extensive and general use and easy application of signs (...) that constitute the true nature of language.[25]

---

[20] Ibid., § 138, 104. "The soul is without composition of parts, one pure simple undivided being. Whatever distinction of faculties or parts we may conceive in it arises only from its various acts or operations about ideas." It is important to note that this passage figures only in the original manuscript of the *Principles*, and not in the final printed version; such an idea is also to be found later in *Alciphron*.

[21] Ibid., §§ 89, 79. "*Thing* or *being* is the most general name of all, it comprehends under it two kinds entirely distinct and heterogeneous, and which have nothing common but the name, to wit, *spirits* and *ideas*."

[22] Ibid., §§ 98, 83. "Time therefore being nothing, abstracted from the succession of ideas in our minds, it follows that the duration of any finite spirit must be estimated by the number of ideas or actions succeeding each other in that same spirit or mind. Hence it is a plain consequence that the soul always thinks."

[23] Berkeley, *Three Dialogues between Hylas and Philonous*, in *Works*, II, 188. "PHILONOUS. – Have all other animals as good grounds to think the same of the figure and extension which they see and fell? HYLAS. – Without doubt, if they have any thought at all."

[24] John Sergeant, too, would find fault with the Lockean conception of the animal as given by Locke in the same passage referred to by Berkeley. See John Sergeant, *Solid Philosophy* (London: Clavil, Roper, and Metcalf, 1697), 1–18. It is possible that Berkeley, having attentively read the book while he was writing his *Philosophical Commentaries*, was inspired by Sergeant.

[25] Berkeley, *Alciphron*, IV, in *Works*, III, 157. This passage does not figure in the two first editions of 1732, but only in the third one of 1752.

To put it in Leibnizian terms, then, we must distinguish between those souls having perception and appetite and those additionally having judgement, which is to say souls having reason. This is confirmed in the first two dialogues of *Alciphron*, in which Crito and Euphranor try to draw a clear line of demarcation between men and beasts when confronted by the free-thinkers Lysiclès and Alciphron who would efface any such difference by nature or degree. Out of the opposition of these two philosophies, idealism versus materialism, the Berkeleian conception of the animal emerges a little more clearly. He views the animal as a living being with a soul that allows it to perceive, move, imagine, and remember, but in no way to reason. The period of its existence is limited by its corporeal death, which would seem to indicate that its soul is mortal. Man, by contrast, is a being naturally endowed with the faculty to think, to desire, and to speak, even if it takes a certain amount of time before these capacities are manifest; and the purposes of man's existence are quite different. He alone is a moral being accountable for his actions. It is with regard to these faculties unique to mankind that Berkeley denounces the stultifying, purely materialistic definition of man the free-thinkers would propose to their contemporaries, according to which man's greatest happiness is to be found in sensual pleasures; Lysicles, for example, says that man is an animal like any other and should therefore heed the same natural directives and let his passions prevail, rather than try to fight them in the name of some artificially imposed asceticism.

But how is one to respond to the free-thinkers when one postulates that the difference between animals and men is merely one of degree? By insisting, as Euphranor does in the second dialogue, on reason being *the* human specificity that introduces inequality into the chain of beings; sensitivity, constituting the essence of the animal soul, is for man but the inferior part of his. If the superior part of his soul is reason, man's finality is then different from an animal's, and the pleasures of his senses can no longer constitute his highest good.

> Man and beast, having different natures, seem to have different faculties, different enjoyments, and different sorts of happiness. You can easily conceive that the sort of life which makes the happiness of a mole or a bat would be a very wretched one for an eagle. And may you not as well conceive that the happiness of a brute can never constitute the true happiness of a man? A beast, without reflexion or remorse, without foresight or appetite of immortality, without notion of vice and virtue, or order, or reason, or knowledge! What motives, what grounds, can there be for bringing down man, in whom are all these things, to a level with such a creature.[26]

Berkeley's problem is that he must fight on two fronts at once. Against the Cartesians he must elevate the animal above being a mere machine stripped of any real perception, feeling, or passions, while against the free-thinkers he must abase the animal to make clear how it differs from human beings. Caught between these two lines of fire, Berkeley's argumentation does not always seem coherent; he grants a soul, but not thought, to animals, yet often defines the soul as what thinks and thinking as the activity of the soul, without specifying each time that he says

[26] Berkeley, *Alciphron*, II, in *Works*, III, 86–87.

this that he is only referring specifically to human souls.[27] By defining the soul or the mind not only by perception but also by thought and action, Berkeley leaves the necessary demarcation between animals and men unspoken, and in this way cultivates errors of interpretation among hurried readers.

If Berkeley had to be taken at his word, then animals, being endowed with minds, or souls, should truly exercise will – should be able to freely determine themselves – with all the moral consequences this implies. But such is certainly not Berkeley's conception. The degree of difference between animals and men is so great that only men can be called truly free. And this fundamental difference, freedom, holds true on the level of passions no less than the level of reason. Animals and men have many passions in common, such as the desire to be happy, which for man in particular translates into the desire to share and contribute to this happiness among others.[28] But man also has passions that are uniquely his own, such as his desire for immortality which is for Berkeley a sign of the endurance of the soul beyond death. Indeed, the order reigning at the physical level might be considered by analogy to reign at the moral level as well: just as the passions of animals are perfectly adapted to the preservation of themselves and of their species, so must the passions of human beings also be adapted to their own purposes, namely the achievement of one's own salvation as well as one's service to the public good. The human desire for immortality thus has its counterpart among animals, divine providence exercising itself at every level and procuring for every being what is proportionate to its desires.[29] If man is internally motivated by the desire to perfect himself and attain "the sublimer pleasures of reason, which discover the causes and designs, the frame, connexion and symmetry of things, and fill the mind with the contemplation of intellectual beauty, order and truth," his desire must then be proportionate to the distinctiveness of his spirituality within the great chain of being, which is namely his ability to grasp the nature of things and thereby deduce the divine principle by which his soul is intended and caused to be immortal.[30]

In addition to this natural desire for immortality, man has other specific passions explainable first and foremost by his status as a rational being. It is because he can free himself from immediate perceptions that man can invent imaginary pleasures and fears for himself that are unknown to animals.[31] This alliance of reason and passion is what makes a creature free, and why the free-thinkers, who downplay the importance of the rational and promote the passions, must be fiercely fought, while ancient idealist philosophy and Christianity, which emphasize the importance of this duality, must be rehabilitated and defended.

---

[27] Berkeley, *Principles*, § 139, in *Works*, II, 105. "A soul or spirit is an active being whose existence consists not in being perceived, but in perceiving ideas and thinking."

[28] On this point, see Berkeley's interesting essay "Happiness" published in the *Guardian*, in *Works*, VII, 214–217.

[29] See Berkeley's two essays "The Future State" and "Immortality" published in the *Guardian*, in *Works*, VII, 181–184 and 222–224.

[30] See Berkeley's essay "Public Schools and Universities" published in the *Guardian*, in *Works*, VII, 203.

[31] On this point, see Berkeley, "Immortality".

> There are two parts in our nature, the baser, which consists of our senses and passions, and the more noble and rational, which is properly the human part, the other being common to us with brutes. The inferior part is generally much stronger, and has always the start of reason, which if in the perpetual struggle between them it were not aided from heaven by religion, would almost universally be vanquished, and man become a slave to his passions, which [is] the most grievous and shameful slavery.[32]

The animal is utterly the slave of its passions, its powers limited to actions based on the singular and immediately present. Berkeley's insistence on the natural order and harmony of things, such that animals' instincts, desires, and passions obey precise laws, is well known. Surprisingly, in making the world a perfect machine ordered to the simplest and most general laws – a machine of which the *Siris* ventures a rough sketch – Berkeley seems to return to a Cartesian conception of the animal, but now as the component of another machine. The analogy between the unconscious bodily movements of humans and the actions of animals, as proposed in Section 257 of the *Siris*, distinctly echoes certain Cartesian reflections which Berkeley had pretended to have surpassed:

> It must be owned, we are not conscious of the systole and diastole of the heart, or the motion of the diaphragm. It may not nevertheless be thence inferred that unknowing nature can act regularly, as well as ourselves. The true inference is that the self-thinking individual, or human person, is not the real author of those natural motions. And, in fact, no man blames himself if they are wrong, or values himself if they are right. The same may be said of the fingers of a musician, which some object to be moved by habit which understands not; it being evident that what is done by rule must proceed from something that understands the rule; therefore, if not from the musician himself, from some other active intelligence, the same perhaps which governs bees and spiders, and moves the limbs of those who walk in their sleep.[33]

To hold this view, however, would entail forgetting that animals perceive perfectly well on their own, have passions of their own, and act according to their own nature, making them automatons of a kind different from Cartesian machines. It would also be to omit the fact that Berkeley does not wish to banish substantial forms from the world, but rather to reintroduce them, as Leibniz did before him, by presenting the world as an immense living organism and animals as its natural automatons. From a slightly different perspective, the *Siris* even makes the soul of the world itself the real cause of worldly actions of which the movements of animals are but occasions.

What remains outstanding is the difficulty, pointed out by Samuel Johnson, that is caused by Berkeley's terminological ambiguity concerning the soul. In accepting the possibility that animals might have souls, the question of the probable immortality of such souls cannot be avoided. It is a question, however, that could only ever have been raised from within a Cartesian philosophical context, and not from a Berkeleian one in which only *rational* souls are immortal. At the time of the *Philosophical Commentaries*, Berkeley would remark rather enigmatically that "the immortality of the soul [is] easily conceived, or rather the immortality of the

[32] Berkeley, "Happiness", in *Works*, VII, 216.

[33] Berkeley, *Siris*, § 257, in *Works*, V, 129. See also *Siris*, §§ 277, 130.

person, that of the soul not being necessary"[34] – a passage which the *Siris* would echo about 30 years later, stating that "personality is the indivisible centre of the soul, or mind, which is a monad so far forth as she is a person. Therefore person is really that which exists, inasmuch as it participates of the divine unity."[35] A rational being's participation in the divine intelligence is what explains the immortality of its soul as a sign of individual personality. As for the souls of plants or animals, it seems more logical to suppose their annihilation with the perishing of the bodies they had animated. After all, the divine omnipotence can create or destroy whatever beings might no longer be suitable to the harmony of the whole as it pleases, whether they be souls or bodies. Ontologically, this costs God nothing, "nothing being more evident than that an omnipotent spirit can indifferently produce every thing by a mere *fiat* or act of his will," and can likewise destroy them.[36]

In conclusion, Berkeley's theory of animality, hearkening in some ways to Leibniz's, does indeed have the coherence to survive the challenges of Baxter and Johnson, if not enough coherence to entirely dissolve all tensions. In his response to Johnson of 24 March 1730, Berkeley recommends to Johnson that he come for a visit to clarify any points on which his philosophy may seem unclear, no doubt with Johnson's questions about the nature of animals in mind. If any such conversation ever did take place, some trace of it must have survived that would give us Berkeley's real position on animality. And there may indeed be such a trace. In his *Ethica* of 1746, published again in his *Elementa Philosophica* of 1752 and dedicated to Berkeley, Johnson would return to what had caused him problems in 1730, and in a few lines of his own provides a conception of the animal no doubt relatively close to Berkeley's:

> Nor yet am I a beast, a horse, a dog, or an ox, etc., for though they appear to see, hear, etc., and to feel pleasure and pain as I do, and can move themselves spontaneously from place to place; yet they have but low, grovelling sensations, exertions and enjoyments. They appear to have no notion of anything but the objects of sense, can conceive nothing of duty and sin, and seem capable of no enjoyment of anything but meat and drink, and the means of continuing their species, and defending themselves; and these only are the things to which their exertions and activity tend. With regard to these, they have, indeed, a wonderful sagacity, and what looks like reasoning, design and contrivance, and a social tendency; but these do not seem to be anything of their own, because they have them originally, and do not acquire them by teaching, trial or industry. This sagacity therefore seems to be what we call an instinct, by which word nothing else can be meant, but that they are rather passively acted and conducted by some other being; some governing mind on whom they depend, according to certain laws of nature which he has established, than that they act from any principle of deliberation and design within themselves.[37]

---

[34] Berkeley, *Philosophical Commentaries*, B14, in *Works*, I, 9.

[35] Berkeley, *Siris*, § 346, in *Works*, V, 156.

[36] Berkeley, *Principles*, § 152, in *Works*, II, 111.

[37] Samuel Johnson, *Ethica*, I, 1, §§ 3–4, in *Elementa philosophica* (Philadelphia: B. Franklin and D. Hall, 1752) 14–15.

Johnson's conception in this passage has the merit of clarifying what could still seem obscure or incoherent in Berkeley with a simplicity that brings to light the real nature of animals and how man's own nature is thereby distinct. But its simplicity is misleading, still not resolving the tensions reviewed in the preceding. The Cartesian conception of the animal, while perhaps "paradoxical" – as Pierre Bayle described it – at least has the merit of resolving such tensions, even if to the detriment of animals.[38]

[38] I want to thank Frank Cameron, Carol Collier, Jeff Hilderley and Syliane Malinowski-Charles for their help in translating this paper.

Index